Praise for
Beer Is Proof God Loves Us

"There is something for everyone in this splendid book on the history, business, and science of beer, enriched with the author's gentle, generous humor and deep insights into the soul of the brewing craft."

—**Xiang S. Yin**, Technology Director,
Cargill Malt, Wayzata, Minnesota

"A hugely entertaining and highly informative romp through the history and complexities of life in general and brewing in particular. Essential refreshment for the beer drinker and thinker and anyone even remotely associated with the brewing industry."

—**Peter Ward**, President,
Institute of Brewing and Distilling (2008-2010)

"A must read for anyone who likes beer, brews beer, or is in or considering joining the brewing industry. Bamforth leads the reader through the industry using anecdotes and stories in a way that is very entertaining."

—**Rob McCaig**, Managing Director and Director of Brewing Technology
for the Canadian Malting Barley Technical Centre,
Winnipeg, Manitoba, Canada

"A very readable volume for people who wish to learn about the brewing process, its history, and evolution. The anecdotes are particularly entertaining and enlightening!"

—**Graham G. Stewart**, Emeritus Professor in Brewing and Distilling,
Heriot-Watt University, Edinburgh, Scotland; Special Professor, University
of Nottingham; and Alcohol Technology Consultant, GGStewart Associates

"Bamforth is without a doubt the world's most knowledgeable beer researcher, writing in a wistfully reflective mode, pondering the traditions, current manifestations, and future of the beverage we love."

—**Dr. Evan Evans**, the Tasmanian Beer Doctor,
University of Tasmania, Hobart,
Tasmania, Australia

"Perhaps the best manifesto defending the value of beer and beer culture, in all its aspects, colors, and flavors. A unique glance into the brain of the beer pope, this brilliant book crams a lifetime of professional experience and knowledge of beer and the beer industry into 250 delightful pages that read like an interesting and witty philosophical autobiography. A must-read."

—**Professor Kevin Verstrepen**,
University of Leuven and VIB, Belgium

Beer Is Proof
God Loves Us

For my growing family.

About the Title

It is now generally believed that, whereas Benjamin Franklin made many great observations, he did not actually say that "beer is proof that God loves us and wants us to be happy." It seems that he did write, in a 1779 letter to the French economist André Morellet: "Behold the rain which descends from heaven upon our vineyards, there it enters the roots of the vines, to be changed into wine, a constant proof that God loves us, and loves to see us happy." I am sure he had beer in his heart of hearts, though.

Contents

Preface

This is not the book that I thought it was going to be.

Some while ago I started writing a book with the word "God" in the title. It wasn't really about beer. It wasn't really about God. It was rather more to do with me. Call it what you will. Midlife crisis? Narcissism? Writing therapy?

Whichever it was, or whether it was something entirely different, it clearly wasn't the right book. And, yet, there was a message in that manuscript that I felt I needed to put into the world.

Which is when Kirk Jensen called. I had worked with him on my first beer book.[1] I told him that I had a manuscript that was fundamentally autobiographical. I said it was part beer, part spirituality. I said I was feeling uncertain about it. He was keen to see what might evolve from the idea.

Which is how we arrived at what you have in your hands. It is indeed a book about beer, albeit perhaps one that comes to the subject from a somewhat unusual, even obtuse angle. And yet, egotistically perhaps, it is also a somewhat personal perspective. To a large extent I have employed endnotes to collect many of these nostalgic ramblings, so that they do not detract from the hoped-for flow of the main text. However, perhaps the perusal of those notes might just strike a chord with the reader. The endnotes are also intended as a repository of other facts, figures, and clarifications (and I see that I have already used my first endnote). I do realize that many people studiously avoid endnotes, but I really do encourage you to read mine, for there is more than the occasional take-home message there. And some of them may even make you smile.

People often ask me how I find the time to write so much.[2] The answer is that, of course, I enjoy it, and that is nine parts of achieving anything. The other reason of course is that I am blessed—not to have talent, but rather to have the most beautiful wife, Diane. I have known her since February 12, 1972, and we have been married since October 9, 1976.[3] She is the heart of our growing family in every respect. Without her I would not be who I am today. She is the one who should really write a book about God.

In writing this book I am grateful to a number of people, not least Kirk Jensen for his steady and forthright guidance. I also acknowledge Larry Nelson, the indefatigable editor of the *Brewers Guardian*, in whose pages over the years I have developed many of the ideas that are built upon in this book.

Introduction

My regular haunt as a boy was a pub called *The Owl* (see Figure 0.1). I was not yet 17, and the legal drinking age in England was (and still is) 18. Friday evenings. One or two pints of Walker's Best Bitter.[1] A bag of crisps (a.k.a. chips) with a tiny blue bag of salt in every pack.[2] And Woodbine cigarettes, of which perhaps three or four would tremble on my lips. I would observe the comings and goings, mostly of the male gender (women then, as now, pleased my eyes more, but in those days they were heavily outnumbered in the pub). Many of the men were tough-as-teak workers, some clad in clogs, leaning against the bar, throwing darts, or rattling dominoes as they took their accustomed places in the dusty oaken furniture solidly set on rustic flooring. No television, no piped music. The food was restricted to pickled eggs, crisps, scratchings,[3] and perhaps the offerings from the basket of the fish man who did his rounds of the pubs, with his cockles, whelks, and mussels.[4] He jockeyed for position with the bonneted Sally Army woman and her *War Cry*.[5]

Figure 0.1   The Owl in Up Holland, with thanks to Sarah Mills.

Arthur Koestler[6] wrote, "When all is said, its atmosphere (England's) still contains fewer germs of aggression and brutality per cubic foot in a crowded bus, pub or queue than in any other country in which I have lived." Not once in the pubs of 1960s Lancashire did I witness anything to contradict this truth.

Who were these men, in their flat caps and overalls, or their simple and well-worn woolen suits? What unfolded in their lives? Were they drinking away their babies' or teenagers' futures, or were they rather savoring precious moments of content amidst the harsh cruelty of their labors? Were they stoking the fire of violence that would afterwards roar through the family home or were they merely rejoicing in bonds of brotherhood with others who knew only too well the rocky roads and unforgiving fields that each of them traversed as laborers and farmers, bricklayers, and quarrymen? This was no less their sanctuary than St Thomas's church[7] or Central

Park, the home of nearby Wigan's prestigious Rugby League team.[8] This was oasis.

And in their glasses would be English ales, nary a lager in sight. Pints (seldom halves) of bitter or mild.[9] The occasional bottle of Jubilee or Mackeson.[10] Perhaps a Bass No. 1 or a Gold Label.[11] Beers with depth and warmth and, yes, nutritional value to complement their impact on conviviality and thirst.

Wigan, immortalized by George Orwell in his *Road to Wigan Pier*,[12] was a few pennies away on a Ribble[13] bus. The pier was a landing stage by the Leeds-Liverpool canal, a place for goods to be offloaded, notably cotton for the mills of the grimy but glorious town. The folks lived in row upon row of small houses, all joined together in grey, damp blocks. Two rooms down and two up and a toilet a freezing trek away down the narrow back yard, with newspaper to clean oneself up and often no light to ensure a satisfactory result. Baths were taken in front of the coal fire in the living room, in a pecking order of father first, mother next, then the children. For those with coal-miner dads it was no treat to be the youngest offspring.

Was it then a wonder that the pub held appeal? Warm, cozy, buzzing with camaraderie and escape.

In England today, pubs are shuttering their doors at a rate of 52 every week. I blame Thatcher, whose ill-judged Beer Laws of the late 1980s led to revered brewers like Bass and Whitbread and Watney selling their breweries to focus on serving the brews of others in spruced-up pubs that are now more restaurant and sports bar than back street boozer. Cleaner, smarter, livelier? Sure. But do they have heart or soul? Yes, they are smoke-free zones,[14] but there are as many folks on the sidewalk outside, spilling into the roadway and littering the pavement with butts and spittle.

Perhaps it is small wonder that many choose no longer to head to the pub and prefer to stay in front of their 70-inch surround-sound televisions, chugging on canned lager bought at fiercely competitive rates from a supermarket chain that commands one in every seven pounds of disposable income in the British Isles and which squeezes the remaining UK brewers to the measliest of margins as they entice the shopper to become solitary suppers of beers with names very different from those of yore.

Beers from breweries like the multinational behemoth Anheuser-Busch InBev, which commands nearly 25 percent of the world's beer market, more than twice as much as the nearest competitor, South African Breweries-Miller. Stella Artois, Budweiser, Becks: all brands owned by the biggest of breweries. Excellent beers, of course, but at what risk to other smaller traditional labels?

The world of beer is hugely different from that I first glimpsed as a too young drinker close to the dark satanic mills[15] of my native Northern England. Has beer, I wonder, lost its soul?

Or is it, rather, me that is the dinosaur? Is the enormous consolidation that has been the hallmark of the world's brewing industry for decades nothing more than business evolution writ large as survival of the fittest? Do the beers that folks enjoy today—and the latter day "near beer" which is the malternative (think *Smirnoff Ice*)—speak to a new age of Kindle, Facebook, and fast food?

In truth, there remains much for this hoary old traditionalist to delight in: the burgeoning craft beer sector in his new motherland, the United States. A growing global realization that beer, rather than wine, is the ideal accompaniment to foods of all types and (whisper it) is actually good for you, in moderation.

All is not lost in the world of beer. Let's go there.

1

Global Concerns

I was on the legendary Fifth Floor of the time-honored St. Louis Brewery of Anheuser-Busch. A dozen or more glasses of Budweiser were before me. Around the table was the cream of the company's corporate brewing staff and me, the newly incumbent Anheuser-Busch Endowed Professor of Malting and Brewing Sciences at the University of California, Davis.[1]

Doug Muhleman, a wonderful Aggie alum[2] and god of matters technical within the august brewing company, invited comments on the beers before us. One by one, the folks around the table proffered their opinion on the samples, which represented the venerable Bud as brewed in all of the locations worldwide where it was produced. In due sequence, my turn arrived. I gulped, thought about my new job title, and said "well, they are all great, all very similar, but this one I find to be a bit sulfury" as I gestured to the lemon-colored liquid in one of the glasses. I needed to demonstrate that I was one smart dude.

A hush fell over the surroundings. I felt all eyes on me. And then I heard someone tapping into his cell phone, as the journey of investigation started into what it was that the esteemed professor had "discovered" in the brew.

I had visions of airline tickets being purchased, jobs being lost, brewers consigned to the Siberia of the company wherever that was (Newark perhaps?). And in an instant I knew that it would be the last time I would pass critical comment in that room. For on the one occasion that I had, with a remark founded on a desire to be perceived as being knowledgeable rather than any genuine ability to find fault with the remarkably consistent product that is Budweiser, the potential impact was too immense to even think about.

There are many people in the United States and beyond who decry Bud. They would be wrong to. For here is a product that, for as long as it has been brewed, which is for rather more than 130 years, has been the ultimate in quality control excellence.[3]

Let there be no confusion here. That a product is gently nuanced in flavor does not make it somehow inferior. The reality is that it is substantially more challenging to consistently make a product of more subtle tone, there being far less opportunity to disguise inconsistency and deterioration than can be the case in a more intensely flavored beverage. And to make such an unswerving beer in numerous locations worldwide, with none but the acutely attuned brewmasters resident in the corporation able to tell one brewery's output apart from another, is a truly astonishing achievement.

✿ ✿ ✿

Doyen of the company from 1975 was August A. Busch III. I recall a former student of mine, newly ensconced at the Fairfield brewery in Northern California, telling me of his first encounter with Mr. Busch. "It was awful," he said. "Mr.

Busch breezed in and spent the whole time firing out questions, challenging and finding fault with pretty much everything that we were doing. Being really critical." I smiled, replying, "You know, that is really a very high class problem. To have a man whose name is on the label showing such interest, commitment, and determination for the best is a wonderful thing. This is someone who will throw money at quality, who believes in being the best. Never knock it. Would you prefer to have a bean counter in corporate headquarters, someone who never comes near the brewery, making decisions solely on the basis of the bottom line and profit margins?"

The stories about August Busch are legion. He is supposed once to have pulled up alongside a Budweiser dray in a midwest city and, noticing that it needed a wash, gave the distributorship five days notice to get their act together or face losing the Bud contract. I am told of the time that a young brewer was summoned to the Busch home to bring some beer for the great man to taste. The youngster duly opened all the beers and placed the bottles in a line alongside sparkling fresh glasses. In came Mr. Busch, took one look at the scene and remonstrated with the young man for throwing away the crown corks from the bottles, for he needed to smell those to make sure that they were not going to be a cause of any flavor taint in the beer.

The same attitudes pervaded the entire company. The commitment to the best started in the barley breeding program of Busch Agricultural Resources in Idaho Falls, Idaho, and the hop development program in the same state and ever onwards through all aspects of the company's operations. The motto in the breweries was "taste, taste, taste." No raw material, no product-in-process, no process stage was excluded from the sampling regime. Brewers would taste teas made of the raw materials, they would taste the water, the sweet wort, the boiled wort, the rinsings from filtering materials, and so on. Nothing (except the caustic used to ensure the pristine cleanliness of the inside of vessels and pipes) was excluded from such organoleptic scrutiny.

Small wonder, then, that the Anheuser-Busch Corporation grew to become the world's leading brewing company in terms of output as well as quality acumen. And yet they could not control everything.

In April 2008 I was a guest at an Anheuser-Busch technical meeting in Scottsdale, Arizona.[4] I was honored to kick off the proceedings with a talk based on my newly published book where I was comparing the worlds of beer and wine.[5] Straight afterwards came a man to the podium from the business operations nerve center in St. Louis. I was reassured to hear him say that Anheuser-Busch was too big to buy when judged against the available dollars that a suitor might have at their disposal. But, in a cautionary afterword, he did stress that the company would never be invulnerable and that it was always prudent to be mindful of size and, therefore, acquisitions should be seriously considered. I knew already that the company had for the most part achieved its magnitude by organic growth, albeit with some additional major investments in China, Mexico, and the United Kingdom.[6]

Less than three months later the aggressive bid of InBev was announced and thus in November 2008 Anheuser-Busch InBev was formed.[7] August Busch III was out.

To search for the root of InBev, we must locate seeds in Belgium and Brazil.

✿ ✿ ✿

The history of beer in Brazil commenced early in the nineteenth century with its import by the Portuguese royal family. It was an expensive commodity, accessible only to the privileged classes, and it was not until 1853 that the first domestic brewery was opened in Rio de Janeiro, producing a brand called Bohemia. In 1885, a group of friends started Companhia Antarctica Paulista in Sao Paulo, at first to sell ice and prepared foods but, not long afterwards, beer. Within five years Antarctica was brewing more than 40,000 hectoliters.[8] Meanwhile in 1888 the Swiss Joseph Villiger began brewing beers in the style of his European roots and named it for the Hindu god, Brahma. As the twentieth century dawned, the substantially grown Antarctica and Brahma began to stretch their hinterland deep into other regions of Brazil, adding breweries and brands, such as Chopp,[9] which enabled the Brahma company to gain ascendancy. Brahma and Antarctica were fierce rivals in both the beer and soft drinks markets. Each grew organically but also through acquisitions as they expanded throughout Brazil. Among the key investments by Brahma was the Skol[10] brand in 1980, a move that soon shifted the company into one of the top ten beer producers worldwide.

Perhaps it was 1990 when the surge of Brahma truly began, with a new chief executive, Marcel Telles, who introduced incentive programs while slashing the payroll and introducing new production and distribution technology. The era of least costs had dawned, as well as global horizons, with Argentina being a first target. For their part, Antarctica was building up their Venezuelan interests. Meanwhile those outside South America were interested in the burgeoning beer business, and thus Brahma made arrangements with Miller to distribute Miller Genuine Draft while Antarctica formed Budweiser Brazil with Anheuser-Busch, while rebuffing a takeover by the US giant. Ironically, when viewed against subsequent events, Antarctica merged at the end of 1999 with Brahma, to produce Companhia de Bebidas das Américas, better known as AmBev, thereby becoming the fourth biggest brewing company in the world, controlling 70 percent of Brazil's beer market, and with expansion plans throughout South America, soon acquiring companies in Uruguay, Paraguay, and undercutting the Quilmes rivals in Argentina to the extent that they too were acquired in 2003. Thus did AmBev control 70 percent of the Argentina beer market, 80 percent in Paraguay, and 55 percent in Uruguay to add to the 70 percent control of the Brazilian business.

✿ ✿ ✿

If the Brazilian beer market is not much more than two centuries old, that in Belgium is rather more long-standing. The Artois brewery, which lends its name to the historic and now global brand Stella Artois (established 1366), was founded in Leuven in the late fourteenth century. Another great brewing company, that of Piedboeuf, was established in 1853. By the 1960s both companies started a three-decade expansion into the Netherlands, France, Italy, and elsewhere in Belgium by acquisitions. They cooperated on the purchase of a third Belgian brewery and, in 1987, merged and hired as CEO José Dedeurwaerder, a Belgian-US joint citizen, to rationalize the operations and deal with organized labor issues. Interbrew, as the company now was known, continued its expansion through acquisition, buying Belgium's Belle-Vue, Hungary's Borsodi Sör, Romania's Bergenbier, and Croatia's Ozujsko.

Interbrew was Europe's fourth largest brewer in the early 1990s, distributing beer in 80 countries. Signs of decline in the European market, however, made the company hierarchy look beyond, and they purchased Canada's John Labatt Ltd. in 1995, the latter company preferring a brewing concern over the Onex Corporation as buyer. Interbrew quickly divested itself of Labatt's nonbeer interests, such as its hockey and baseball clubs. At a stroke, Interbrew gained an extensive North American distribution system that could now ship products such as Stella Artois and Hoegaarden. It brought, too, a 22 percent interest in Mexico's Dos Equis brand as well as the iconic Rolling Rock.

Interbrew began exporting Stella Artois to China via joint ventures, recognizing the world's fastest-growing beer market, while continuing doubts about the European market led to it rationalizing some of its European interests, such as Italy's Moretti, sold to Heineken. However, Interbrew built major stakes in breweries in Bulgaria, Ukraine, Russia, Bosnia, Ukraine, Slovenia, and Germany, such that by 2000 it operated in 23 countries and was number three worldwide, behind Anheuser-Busch and Heineken.

Interbrew's next two major acquisitions were Bass from the UK and Beck's in Germany. As we see in Chapter 2, "The Not-So-Slow Death of a Beer Culture," Margaret Thatcher had severe misgivings about what she perceived to be a monopoly scenario in the UK and very rapidly a number of major brewing companies came into the market. Bass enjoyed 25 percent of the British market, and competitor Whitbread had almost 16 percent. Both companies went on the market in 2000 as Interbrew declared its intention to go public. By June, Interbrew had bought the breweries and brands of both Whitbread and Bass (the British companies themselves survived as hotel and retailing concerns), although the perception that this huge inroad into the UK industry would also constitute a monopoly situation led to Interbrew divesting itself of Bass's major brand Carling Black Label and the breweries that brewed it to Coors. Even then, Interbrew had 20 percent of the British beer business.

The public listing of Interbrew shares now made cash available for further international acquisitions, and Beck's was first. Rumors were that the next purchase would be South African Breweries, but that company itself was intent on globalization, shifting its headquarters to London, and purchasing the likes of Pilsner Urquell in the Czech Republic and Miller from Philip Morris, thereby becoming SAB-Miller, the second biggest brewing company on the planet.

On March 3, 2004, Interbrew and AmBev merged into a single company named InBev, at a stroke giving it a 14 percent share of the global beer business, with interests in 140 countries and making it the world's number one, pushing Anheuser-Busch into second place. And on November 18, 2008, the acquisition of Anheuser-Busch by InBev closed at an inconceivable $52 billion, creating one of the top five consumer products companies in the world and a company producing around 400 million hectoliters of beer annually, with the next biggest competitor, SAB-Miller, standing at 210 million hectoliters.

✿ ✿ ✿

As 2009 dawned, Anheuser-Busch InBev announced the closure of the Stag Brewery in Mortlake, London, with the loss of 182 jobs. Anyone who has watched the Oxford-Cambridge boat race will know of it, right there by the River Thames, close to the finishing line. Rationalization. And what stories that brewery can tell about brewery history and the march of the megabreweries.

The brewery dates from 1487 when it was associated with a monastery. By 1765 it had become a major common brewer[11] and a century later was rebuilt as the 100-acre site that would be bought by Watney in the 1890s and would go on to be a primary brewery for the production of the reviled Red Barrel.[12] Watney's became part of the Grand Metropolitan leisure group and was soon brewing Germany's Holsten and Australia's Foster's under license. Come Thatcher (see Chapter 2), Watney's sold all its plants, including Stag, to Courage, which in turn became part of Scottish & Newcastle, who leased the Mortlake brewery to Anheuser-Busch for the brewing of Budweiser. Scottish & Newcastle became the last of the "big six" British brewers to survive PMT (Post-Margaret Thatcher) and sold out to a Heineken and Carlsberg joint assault, the latter two dividing up the company between them.[13]

Thus did the Stag Brewery find itself vulnerable within the new Anheuser-Busch InBev giantopoly. Result: More than 520 years consigned to the history books and a prime piece of real estate available for regeneration.

❀ ❀ ❀

It was ever thus. Brewing companies have been bought and sold for generations. Take, for instance, the Bass company that was acquired by Interbrew and then rent asunder in the Coors deal.

The monks started brewing in Burton-on-Trent in the twelfth century. Among the commercial brewers that would make the East Midlands town truly famous, surely the "big cheese" was William Bass who started his operation on High Street in 1777 after previously being a transporter of beer for Benjamin Printon. Bass shot to international fame in 1821 with its famed East India Pale Ale, shipped to the Raj.[14] By 1837, the company had become Bass, Ratcliff & Gretton, reflecting the partnership of Bass's grandson with John Gretton and Richard Ratcliff. As the railways expanded, so did the fame and hinterland of the company, and by 1860 the brewery was churning out more than 400,000 barrels a year. There were some 30 or more brewing competitors in the town, but Bass became Britain's biggest brewing company. The popularity of its bottled ale obliged the company to become the first firm to use the Trade Marks Registration Act of 1875 with the registration of the red triangle emblem.[15]

In 1926, the company bought another Burton brewer with a countrywide reputation, Worthington & Company Ltd. A year later, the company bought Thomas Salt's brewery, and six years later, that of James Eadie. But the company, under the chairmanship of Lord Gretton, who was seemingly somewhat stubbornly resistant to change and distracted by a political career,[16] did not embrace the change that it might have done, in particular not buying into tied public houses for the selling of its beer. It was Arthur Manners, assuming the chairmanship in 1947, who drove the company forward in a more businesslike way. Bass acquired holdings in William Hancock & Company and Wenlock Brewery Company. Soon there were 17 subsidiaries throughout the British Isles.

In 1961, then-chairman Sir James Grigg, who had been in Winston Churchill's government cabinet,[17] merged Bass, Ratcliff & Gretton with Birmingham's Mitchells & Butler, a company that itself had grown through acquisitions and which had ruthlessly rationalized production operations, but most importantly had rejoiced in a strong tied house portfolio. This was followed with the merging in 1967 with London-based Charrington (founded 11 years before Bass), with Sheffield's William Stones Ltd. coming under the umbrella a year later. And Hewitt's of Grimsby was snaffled in 1969. So it was now a case of Bass, Mitchells & Butler and Bass Charrington in different regions of the country.

The most critical aspect of the Charrington move was that it had previously merged with United Breweries, owners in the UK of the rights to the Canadian Carling Black Label brand, which would go on to become by far and away Bass's biggest beer.[18] Under ruthless chairman Alan Walker there followed tremendous rationalization as breweries were closed and production consolidated in strategic locations. And the company now had a huge estate of tied houses, to go along-side growing interest in hotels, betting shops, and other leisure activities. By the end of the century, with Margaret Thatcher's Beer Laws that we will visit in the next chapter, the hotels (notably Holiday Inns) became the focus—and Bass as a brewing legend died. The cask Bass brand[19] is these days owned by Anheuser-Busch InBev and is brewed under license in Marston's—a brewery in Burton since 1834[20] and thus a longtime competitor of Bass.

✿ ✿ ✿

So what of this consolidation, ancient and modern? Does it represent nothing more than an incessant quest for domination, profits, and shareholder satisfaction, with the invariable reduction of choice and quality in the products available to the customer? Or is it an unavoidable consequence of economic reality (survival and growth of the fittest) and might it even benefit the consumer?

Consolidation and growth invariably lead to a reduction in employment, as a consequence of the pooling of production into fewer, larger, strategically placed breweries with the closing of inefficient, highly staffed smaller locations. Furthermore, advances in sensor and control technology mean that breweries are increasingly automated: Go into even the largest of breweries and you will see very few employees, with the greatest numbers to be found in packaging, warehousing, and distribution. As can be seen in Figure 1.1, a major component of the cost of a bottle of beer is personnel in production (including packaging). How much more efficient, for example, to have one 2,000-hectoliter fermenter as opposed to ten vessels of 200 hectoliters. The latter are unavoidably less efficient as they individually need to be filled, monitored, emptied, and cleaned.

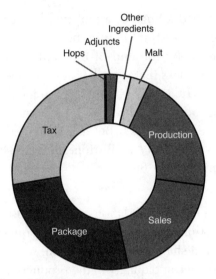

Figure 1.1   The costs within a bottle of beer.

Responsible brewing companies enter into consolidation issues with their eyes and minds open from a technical perspective. (I wonder, however, quite what their hearts are doing, should they give pause for thought about the humanitarian issues surrounding job losses and the inevitable slicing at the heart of local communities when a major employer center is lost.)

Consider, for instance, the act of changing the type of fermenter. A perfect example was given by the shift within Bass during the early eighties from the Burton Union system[21] to cylindro-conical vessels[22] for the fermentation of the legendary Bass Ale. This was not a consequence of any takeover activity, merely the desire of the company to move away from a traditional mode of beer production, one that is more labor intensive and associated with a greater spoilage rate, to a more modern, streamlined, and controllable approach. A reputable company only makes such a move after a very large number of trials, in which process variables are tweaked to ensure that, at the end of the day, there is no impact on smell or taste or any other manifestation of product quality. The changeover in fermentation approach predated me at Bass, but the variables that they played with must have included fermentation temperature, wort composition, and the amount of oxygen supplied to the yeast.[23] I know—because Bass's mentality as regards quality was identical to that of August A. Busch III— they would have ensured that the product "match" was perfect. And yet, inevitably, when it became known that the change had been made, there were the draught Bass aficionados who insisted that the product was "not a patch on what it was before they started buggering about with it." Perception becomes nine parts of reality.

Brewers particularly run into this type of problem when they acquire companies with very different technology or when they seek to have their beers brewed under franchise by companies with alternate philosophies when it comes to beer production.[24] I very quickly learned when I was Director of Research at BRF International[25] and trying to identify research projects that would satisfy all my customers, that brewers quickly become adherents to favored brewing approaches. Perhaps the most strident are the Germans, but they are not alone. Some insist on "bright worts," others on "dirty worts."[26] Related to this, some prefer lauter tuns, others mash filters.[27] There are those who use horizontal fermenters, others vertical ones.[28] The list goes on. And each and every one of these differences impacts the flavor of the beer. However, by changing parameters of the type referred to previously, so can the differences be eliminated. It truly is possible to produce wonderfully matched beers in widely divergent breweries.

It is also axiomatic that recognition be taken of the importance of the raw materials. The correct yeast strain must be used (although this can be debated for some of the more strongly malty and hoppy brews[29]). The malt and hops must be within the declared specification; not least they must be of the declared variety. And the water must be right.

Much is said about the importance of water in brewing. Rightly so, for most beers are at least 90 percent water. The reality is that technology is such that the water specified for the brewing of any beer anywhere in the world can be produced very straightforwardly.[30] To make the very soft water prized in Pilsen involves simple filtration technology to remove salts. By adding calcium salts one can easily make water to match the very hard stuff from Burton-on-Trent—heck, the Germans even have a word for it ("Burtonization"). Rocky Mountain water is a charming concept (and I love the folks in the Golden brewery[31]—they are smart, capable, and fun), but that water is not magical. I can make it right here in Davis.

An old boss of mine (a chemical engineer and therefore coldly logical) once described beer as being "slightly contaminated water." I would contend that if such it is, then it is an awesome form of impurity, but nonetheless the observation does speak to the fact that beer is an extremely aqueous commodity. That being the case, it simply does not make sense to ship it vast distances. It is so much more sensible to brew as close to the drinker as possible; therefore the concept of franchise brewing.

The other reality is that of beer's inherent instability. There is more than a grain of truth in the adage that beer is never better than when first brewed and when drunk close to the brewery. For the majority of beers it is downhill from the moment that the crown cork goes on the bottle, the lid goes on the can, or the keg is racked. Beer is susceptible to a number of changes; the most challenging of all being staling. In Chapter 5, "So What Is Quality?," I discuss this issue, which spills into matters philosophical and psychological, even physiological. And indeed there are a very few beers, notably those of very high alcohol content, that may actually benefit from storage.[32] But for the vast majority of beers there will be a progressive development of cardboard, wet paper, dog pee, straw, and other aroma notes that I, at least, find reprehensible, characteristics that detract from drinkability.

This issue of flavor instability is highly pertinent in consideration of the globalization of the beer market and the growth of the mighty brewers. On the one hand, these brewers certainly should (and often do) have better control over the key agent that causes the flavor deterioration of beer, namely oxygen. They have invested in the latest in packaging lines that minimize air levels. They can afford the most accurate oxygen-measuring equipment and the systems to put in place to respond to it. And in theory at least, by brewing in plants local to the consumer base, they are able to deliver younger beer than would be the case if they were exporting their products. As we have seen, as long as the raw materials and processes are specified and controlled, it is entirely possible to re-create any brand in any brewery in the world (see my earlier Budweiser experience). Nonetheless, there are plenty of instances of major brands continuing to be exported to markets many thousands of miles from home base, taking advantage of the cachet of a certain provenance. Heineken, Guinness, Bass, and Corona are examples of imported brands in the USA that each speak to a national heritage, respectively Holland, Ireland, England, and Mexico. The US drinker seems to prize the import imprint, despite inevitable aged character in the products.[33, 34]

The bigger the company, the bigger the marketing strengths it possesses. And so brands such as Corona, practically unheard of in the US 25 years ago, have reached huge volumes very much on a platform of a trendy beverage from south of the border: the flint glass bottle, the slice of lime, with images of gently rolling surf, wide sandy beaches, and beautifully bronzed bodies. Silence to be savored. The risk, as companies get ever bigger, is that such marketing-forced consumerism will lead to a rationalization of brands and the loss of esteemed beers that are simply beyond the numbers capable of being handled efficiently, whether from a production and packaging, distribution, or promotional perspective. If we consider Anheuser-Busch InBev, for instance, then at the last count it owned more than 300 brands, from Bud to Boddingtons, Harbin to Hoegaarden, Michelob to Murphy's, and Spaten to St. Pauli Girl. One must wonder how many of these products will still be extant 10 or 20 years from now. There is already an approach in this (and many other) brewing companies to developing numerous new beers, trying them in the marketplace, and quickly withdrawing all but the most successful.[35] But there are also brands of much longer standing that seem to be hot potatoes.

Take for instance Rolling Rock. Let's shoot back to 1893 and the founding of the Latrobe Brewing Company in the tiny town in the foothills of the Allegheny Mountains in Pennsylvania. The locals reckon that it was a local enclave of Benedictine monks that first did the brewing, but with rather more certainty we can say that the company was victim of Volstead,[36] and the brewery closed as Prohibition was enacted. Under new owners, the Tito family, the brewery reopened in 1933, with two beers called Latrobe Old German and Latrobe Pilsner. Six years later, though, they launched the beer that made Latrobe famous: Rolling Rock, named in reflection of the river with its smooth pebbles that supplied water to the brewery and packaged in a green glass bottle bearing a horse head-and-steeplechase icon that to this day renders the brand unmistakable on retail shelving. The beer was barely marketed, yet fetched an intensely loyal following in southwest Pennsylvania as well as a presence in several states in the northeast. In 1974, 720,000 barrels of Rolling Rock were produced. As other companies aggressively promoted their brands, Latrobe held back, and volumes of Rolling Rock declined significantly. The Titos sold the company in 1985 to a buyout concern called the Sundor Group, which sought to turn around the business prior to a resale. Sundor boosted marketing strategies but throttled back on capital investment: a classic conflict between going gung-ho on sales, while jeopardizing the quality of the very product on offer. Two years after Sundor came in, it sold Latrobe to Labatt. Now the Rolling Rock brand was in the hands of a company that totally respected quality but also possessed a keen eye for marketing (witness its original concept of Ice Beer[37]). Indeed Labatt made a big play of the mysterious number 33 long since found on the green bottle, and this came at the heart of the marketing strategy. The June 20, 1994, issue of *Brandweek* gave the Labatt marketing man John Chappell's description of Rolling Rock as being "A natural, high-quality beer with an easy, genuine charm that comes from the Rolling Rock name and the traditional, small-town Latrobe Brewery that uses the mountain spring water in special green bottles." The sentence contained 33 words—by accident or to encourage brand devotees to come up with their own theories for the origin of the number? Whatever the reason, Rolling Rock was rolled out around the United States, and by the early 1990s the Latrobe brewery (which was attracting investment from Labatt) was churning out more than 1 million barrels per annum. And the product could be now marketed at a higher price.

Interbrew, since 1995 owners of Labatt and therefore
Latrobe, seemed committed to the Rolling Rock brand. In
2000 the declared intention was to double production capacity
with the expenditure of $14.5 million on a new packaging line.
But what is constant in this world? In May 2006 the new
InBev company decided it could offload the brand—and duly
sold it for $82 million to Anheuser-Busch, subsequently sell-
ing the brewery to the company in La Crosse, Wisconsin, that
runs the old Heileman brewery.[38] The good folks of Pennsylvania were up in arms: How could Rolling Rock possibly be
brewed by Anheuser-Busch, especially anywhere other than
by the Latrobe River? I had a different question of my friend,
Doug, the chief technical officer at Anheuser-Busch. For I
knew as well as he did that the overwhelming characteristic of
Rolling Rock is a dimethyl sulfide (DMS) note[39] that most
brewers consider a serious defect when present at the levels to
be found in Rolling Rock. I remember offering "I guess you
will gradually lower the DMS level over a period of time, so
that nobody will notice that the product is changing." "No,
Charlie," Doug replied, "we will learn to brew a defect." And
they did, faithfully adhering to the recipe that they had inher-
ited and sticking to the principle of delivering to the customer
what the customer expects. In fact, knowing Anheuser-Busch,
the product would, batch-to-batch, be more consistently
adherent to its recipe and provenance than would have been
the case prior to the acquisition.

With what irony, then, was the brand restored to the
InBev portfolio with the acquisition by the latter of Anheuser-
Busch. And so no surprise to read the *Wall Street Journal*
article on April 13, 2009, saying that "Brewing giant
Anheuser-Busch InBev is exploring the sale of its storied but
struggling Rolling Rock brand, according to people familiar
with the matter." The article went on to say, "When Anheuser

bought Rolling Rock in 2006, it sought to reposition the brand to compete in the fast-expanding, small-batch 'craft' beer segment. But sales, which already were declining under InBev, have continued to wane. Last year, Rolling Rock sales slipped 13 percent from a year earlier in volume terms to 7.4 million cases, according to Beverage Information Group, a market-research firm in Norwalk, Conn. In 2004, Rolling Rock sold nearly 11 million cases."

As I say, when push comes to shove, there are only so many brands that a company can handle.

✿ ✿ ✿

Every year the indefatigable Emeritus professor Michael Lewis and I generate new recruits eager for brewing pastures in companies large and small. In 2010 there were 66 students in my main brewing class on campus, 32 in the practical brewing class, and 40 in the extension class.[40] Not all of the campus students aspire to the brewing industry (some set their sights rather lower—winemaking, for instance), but all those in the Extension class are already in the industry, or the greater number aspire to be.

Over the years there has been a gratifying flow of Davis graduates into the brewing industry. I fear for the future. As companies consolidate, the most recent example being Miller and Coors in the US,[41] it can only mean fewer openings. Many of the students, it must be said, are intent on the craft sector, however mistakenly regarding the big guys as corporate America, and some of them naively buying into the notion of "industrial beer."[42] The reality is that a rather more comfortable living, founded on the greater range of career-advancement opportunities, can be had in the "majors." The smaller companies do not generally pay well: Some appeal to one's passion to be hands-on in all aspects of the operation, allied of course to copious free beer and the opportunity to converse with the customer—"Hi, I'm Jack. I brewed this beer!"

Those joining the big guys need to recognize three things. First, the need is for brewers (more strictly speaking, brewery managers) in all of their locations, even the less sexy places. There is a big world outside California. Second, the candidate must never, ever have had a DUI.[43] Brewers need to be genuine role models for responsibility.[44] And, third, the company will be snipping a hair sample to check for any interesting social activity.[45] I swear that some of my students fail on at least two of the three, although to the best of my knowledge nobody has been rejected on all counts.

And so the reservoir of talent seems to be very full right now. While there is a trickle going to replace retirements and feed the gratifyingly growing craft sector, there is inevitable seepage for want of openings. I look to my conscience: Can we hand-on-heart continue to encourage all those who want to live their dreams through becoming brewers?

I was dismayed to hear a little while back of one chief executive saying that only a tiny proportion of his employees really mattered to him, because they represented the difference between success and failure. It straightway put me in mind of my old boss, Robin Manners, chief executive of Bass Brewers and grandson of the company's erstwhile chairman. He said to me one day, "Two things matter to this company, Charlie: One is people, and the other is quality. And if you look after the people, they will ensure the quality." What a contrast.

It is fashionable to talk of a War for Talent, the argument being that really worthwhile recruits that will move a company forward are thin on the ground. I rather think that there is ample human resource available, either already employed in the brewing industry, located in other industries, or emerging through the academy. And I know of far too many excellent people downsized from the world's major brewing companies, surely a consequence of companies ripping out expense to present themselves as least-cost operators, thereby impressing the stock markets as they join the fight to get their products a competitive edge on the shelves of equally ruthless supermarkets. If only companies of all shapes and sizes considered employee and customer alike as an individual human being in a nurturing environment.

Buddhists speak of loving kindness. I think that is what my old boss, Robin, was really referring to: Treating everyone from the main board to the janitor as equally deserving of respect and regard for their contribution to the whole—and that esteem and goodwill extended to suppliers on the one hand and to the beer market on the other. We at Bass were simply great guys to deal with, and that counted for a great deal and made us the most successful company in our market. That is, we were, until the bean counters arrived.[46]

It is far too easy for in-your-face business sultans to scream the old adage that nothing is as inevitable as change and that it is only through change that success can be achieved. The simple reality is that business decisions, especially in publicly owned companies, are made on the basis of the bottom line and no consideration of tradition or status quo, unless it satisfies a marketing strategy. In relation to this, ponder for a moment Pabst Blue Ribbon, once the quintessential blue-collar low-cost beer. These days it is trendy, maybe even sexy for all I know, to be seen with a can of PBR. It is not for any rediscovered uniqueness about the brand. It is because "retro" sells. The traditionalists of course might legitimately argue that it would have been better still if the original brewers[47] of PBR had never been subsumed in the first place.

As we have seen from Figure 1.1, a huge slug of the cost of
a bottle of beer goes to sales and marketing. Without doubt, a
customer needs to be receptive, and no matter how catchy the
advertising, a beer that is intrinsically wrong will not sell.[48] Yet
nobody can mistake the power of persuasion and the ability of
marketing (allied to technical advances) to shift drinking preferences. Thus we have, for instance, the baffling (at least to
me) surge toward the iciest of lagers in soggy old England, a
nation generally believed to cling to "warm" ale.[49]

It would be very easy for me to be perceived as a dinosaur,
yearning for a better time much as a bicyclist[50] might resent
the advent of Maserati. Yet the contemplative and meditative
me savors what I like to call the Slow Beer Movement: Traditional brewers with long-standing names and values brewing
beers of heritage and culture, rather than fast beers of short
lifetime and dubious provenance that search out the lowest
common denominator. I even hear that within one company
the management don't speak of beer, but rather call it "liquid."

And so I applaud the craft sector, though even here the
Hyde of extreme brewing (ludicrous hopping rates, bizarre
ingredients) all too frequently escapes the common-sense
calm and beauty of Jekyllian values. We go there in Chapter 4,
"On the Other Hand: The Rebirth of a Beer Ethos." Let us
first, however, head back to my heritage.

2

The Not-So-Slow Death of a Beer Culture

They had beer-and-sandwich lunches, so we might have hoped for a better outcome. The reality is that those meetings between the British government of the 1970s and the mighty unions, of most note being Arthur Scargill's National Union of Mineworkers, were a futile attempt to find common ground. One likes to think that there was a genuine attempt on the part of Edward Heath's Tory government[1] and the rabid socialism of the likes of Scargill to find a deal that would not bankrupt the government while confirming the union workers in worthwhile, safe, and sufficiently rewarding employment. The reality was that we had abject weakness on the one hand and seeming disingenuous attitudes on the other.[2]

Margaret Thatcher had no truck with such approaches. The Iron Lady brought down her mighty fist and the unions were splattered. And so she will go down in history as the Prime Minister who took the nation out of the horrors of strike after strike that paralyzed the country, introducing a market-driven system where rewards were to be had by genuine endeavor and not as a God-given right.

But she got it dead wrong with the brewing industry with the beer orders of 1988, which led directly to the loss of some of the great brewing names. Names like Bass.

I joined Bass, in Burton-on-Trent, in the spring of 1983. It was either that or a place on the faculty at the renowned Imperial College, London: a no-brainer.[3] To join the company that was the UK's equivalent of Anheuser-Busch (albeit writ smaller in volume terms) with its dedication to people and quality was immense. With what pride did we wear the red triangle. We had 13 breweries and major brands such as Carling Black Label, Bass Ale, Tennent's, and Stones Bitter. Wherever you looked in the company there were top people: some of the most gifted brewers and batteries of outstanding quality control folks, through to wonderful staff serving the beer in the pubs.

All was not rosy, though, as I found when they shipped me to the Preston Brook brewery in 1988 to have the "smooth edges knocked off—you must be seen to escape from the Ivory Tower, Charlie." At the root of the problem were the unions that Margaret Thatcher was intent on controlling. The reality at Preston Brook was that control was very much in the hands of the Transport and General Workers Union (TGWU) and the Amalgamated Union of Engineering Workers (AUEW). I certainly learned a lot about human nature in my 2+ year posting at the brewery that some called "Pissed and Broke" and was baffled at how the union concept, founded on the finest principles of support and help for the downtrodden in more draconian days, could be so abused.

Bass had visions of the brewery, close to Runcorn, being a behemoth plant supplying vast swathes of the beer-drinking hinterland owing to its strategic location at the crossroads of north-south and east-west thoroughfares.[4] They made two naïve moves. The first was to entrust the design and construction of the brewery (completed in 1974) to engineers with no feel or empathy with the very special demands that a brewery operation must have. In short they designed a chemical works, badly laid out for all considerations of beer quality. The second was that they had not realized the militancy of the very people that they were recruiting to the workforce: hardened unionists who had honed their rebellious, almost anarchic mentality in the docks of Liverpool.

I was Quality Assurance Manager and, as such, had no members of the TGWU or AUEW in my team of more than 30.[5] But I had plenty to say in management meetings about my perception of what needed to be done to solve our problems. Came the day, then, that the managing director Mike Myers (no relation—this one was no comedian) decided that I should sit in on one of the management-union meetings, so that I could see for myself just how they were dealing with the situation.

Myers, a chain-smoker who was so laid back that he seemed horizontal, started the meeting.

"I would like to kick off by showing you a graph that summarizes the quality performance of Carling Black Label from Preston Brook as compared to the other breweries in the group. As you can see, we have moved from being last to now having the best scores in the Bass organization. And this is a tribute to the Head Brewer, Neil Talbot, and the Quality Assurance Manager, Charlie Bamforth."

The TGWU shop steward interrupted.

"Err, Mike..."

"Yes, Jack."

"Who the f*** is Charlie Bamforth?"

"He's sitting right there," came Myers' reply, as he gestured to me.

The unionite fixed me with his meanest stare, and his eyes narrowed menacingly.

"Never seen you before. You must have a f***ing good hiding place. Must let us know where it is sometime."

Then he instantly swiveled back to glare at Mike Myers.

"But never mind him. It's you, yer ****. You're the one we want."

And thus did a couple of hours unfold in which the union men verbally abused one member of the management team after another (present company excepted—after all, they had no interest in me, as they had said).

I was astounded. It seemed to go on for an eternity as I watched and wondered. Eventually the abuse ended and the unions stood en masse and marched out of the door, slamming it behind them. Mike lit up another cigarette.

"Well," he said "I think that went pretty well."

If I did not know it already from my memories of strikes, disruptions, three-day weeks, and the other miseries owing to the radical left, I certainly now appreciated what Thatcher was trying to do. The brewery by the Mersey was unmanageable. And the pleas of the likes of Talbot and me for investment so that we could properly control quality were doomed to failure at HQ, who would not invest in a brewery that was so seriously dysfunctional.[6] A year after I left the company in 1991 the brewery closed, and the tanks were packed in crates and shipped to communist Eastern Europe.

By that time Thatcher had embarked on what was to me a less justifiable goal than bringing the unions to heel: She set the brewing industry itself in her sights. She decided it was a monopolistic situation.

The UK brewing industry at the time was dominated by The Big Six, namely Bass, Allied, Whitbread, Grand Metropolitan (the erstwhile Watney's), Courage, and Scottish & Newcastle. National companies all. But there were plenty of other, smaller brewing companies, such as Greenall Whitley, Wolverhampton & Dudley, Marston's, Greene King, Shepherd Neame, and many, many more. Every region had its own brewing companies, from St. Austell Breweries and Redruth Breweries in Cornwall to Sinclair Breweries in Orkney.

The wonderful feature of the beer world in Britain at the time was the pub. Although there were "free houses" not tied to a given brewing company, most pubs were owned by the brewers themselves and primarily sold the products of that company, perhaps with the odd bottle of beer from other suppliers as well as the inevitable tap delivering Guinness. It was this vertical integration that stuck in Thatcher's craw: In short she said it was a monopolistic scenario. She would not countenance the argument that in most places there were substantial numbers of pubs owned by different brewing companies. No, the situation must change. She was adamant. In due course (1988) "The Beer Orders" appeared, a.k.a. *The Supply of Beer: A Report on the Supply of Beer for Retail Sale in the United Kingdom.*

In the Monopolies and Mergers Commission report, they themselves pointed out that there were more than 200 brewers in the UK, categorized as

-the 6 national brewers accounting for 75 percent of beer production, 74 percent of the brewer-owned retail estate, and 86 percent of loan ties.

- 11 regional brewers accounting for 11 percent of beer production, 15 percent of the brewer-owned estate, and 8 percent of loan ties.

- 41 local brewers, accounting for about 6 percent of beer production, 10 percent of the estate, and 4 percent of loan ties.

- 3 brewers (namely Carlsberg, Guinness, and Northern Clubs Federation) without tied estate and supplying about 8 percent of beer production and accounting for more than 1 percent of loan ties.

- 160 other brewers, all operating on a very small scale.

Not an inconsiderable diversity, one might argue—compare and contrast for example the situation at the time in the United States where there were far fewer brewing companies, and overwhelming control by a very few.

At the time the Brewers Society,[7] which represented the collective interest of the UK brewers, made the following observations about the brewing industry and the merits of the extant system:

There is self-evident competition between individual public houses locally and therefore between brewers of every size;

- public houses individually and collectively compete for consumers' leisure expenditure, and therefore compete also with clubs, restaurants, home entertainment and many other leisure activities;

- vertical integration in beer retailing has been approved by the European Commission under Regulation 1984/83;

- vertical integration permits close control of product quality from producer through to consumer;

- vertical integration has permitted, and competition has encouraged, brewers to achieve beneficial cost improvements right through the supply chain;

- the brewing industry has an excellent record of product innovation and amenity improvement made possible only by the operating and financial security which vertical integration confers;

- the structure of the industry has, through the tenancy system, allowed thousands of small entrepreneurs to get into business on their own account with minimal capital outlay; and

- loan-tying helps the low-throughput public house and club to start up and stay in business.

In short, The Brewers' Society saw

no widespread or focused evidence of consumer complaint. The industry had adapted readily and well to changing consumer tastes and social trends. The British public house was an important institution, arguably unique in the world, and offering the consumer better value for money than bars in other

countries. Neither the public house, nor the arrange-
ments which kept it as we know it, should be tampered
with.

The Monopolies folks somewhat patronizingly observed,
"There is no doubt in our minds that the Society is formi-
dably effective in championing its members' interests,"
and then proceeded to give their response:

Eloquently though the industry's case has been put,
we are not persuaded that all is well. We have con-
firmed our provisional finding that a complex monop-
oly situation exists in favor of the brewers with tied
estates and loan ties.

This complex monopoly restricts competition at all
levels. Brewers are protected from competition in sup-
plying their managed and tenanted estates because
other brewers do not have access to them. Even in the
free trade many brewers prefer to compete by offering
low-interest loans, which then tie the outlet to them,
rather than by offering beer at lower prices. Whole-
sale prices are higher than they would be in the
absence of the tie. This inevitably feeds through into
high retail prices.

The ownership and loan ties also give little opportu-
nity for an independent wholesaling sector to prosper
and offer competition to the brewers' wholesaling
activities, for example by offering a mix of products
from different producers.

In summary:

- the price of a pint of beer in a public house has risen
too fast in the last few years;

- the high price of lager is not justified by the cost of producing it;

- the variation in wholesale prices between regions of the country is excessive;

- consumer choice is restricted because one brewer does not usually allow another brewer's beer to be sold in the outlets which he owns: this restriction often happens in loan-tied outlets as well;

- consumer choice is further restricted because of brewers' efforts to ensure that their own brands of cider and soft drinks are sold in their outlets;

- tenants are unable to play a full part in meeting consumer preferences, both because of the tie and because the tenant's bargaining position is so much weaker than his landlord's; and

- independent manufacturers and wholesalers of beer and other drinks are allowed only limited access to the on-licensed market.

In summary, we believe that the complex monopoly has enabled brewers with tied estates to frustrate the growth of brewers without tied estates; to do the same to independent wholesalers and manufacturers of cider and soft drinks; to keep tenants in a poor bargaining position; and to stop a strong independent sector emerging to challenge them at the retail level. We believe also that, over time, the monopoly has served to keep the bigger brewers big and the smaller brewers small.

These are serious public interest detriments. Since significant growth in the number of full on-licenses issued is unlikely, we believe that structural changes are essential to secure a more competitive regime which will in turn remedy the detriments.

And then the kill (literally as it turned out), in the shape of their recommendations:

> - *In present circumstances, if the tie (ownership of pubs by breweries) were to be abolished altogether we believe that many regional and local brewers would withdraw from brewing, concentrate on retailing, and leave the market to domination by national and international brand owners. We therefore recommend a ceiling of 2,000 on the number of on-licensed premises which any brewing company or group may own. This ceiling will require the divestment of some 22,000 premises by United Kingdom national brewers. We are recommending a maximum of three years for the divestments to take place.*

> - *in order to improve the market opportunity in the tenanted trade, we recommend that a tenant should be allowed to purchase a minimum of one brand of draught beer from a supplier other than his landlord. We also recommend that there should be no tie whatever for non-alcohol or low-alcohol beers, nor for wines, spirits, ciders, soft drinks or mineral waters.*

And then the astounding observation (in view of what would transpire):

> *If no changes are made we believe it is inevitable that a very small number of brewers will increasingly dominate the supply of beer in the United Kingdom.*

The UK brewing industry was aghast—but the most they could wrangle out of the government before the ensuing bill was passed to law was that they would be allowed to keep not only 2,000 pubs, but also 50 percent of the remaining estate that they may own. Thus, in the case of Bass, owning 6,300 pubs, it was necessary to sell 2,150 outlets.

To summarize the impact, it was quickly realized that there were two ways to go for the big guys: either brew beer or sell it. The retail profit on beer in the pubs is substantially larger than is the wholesale margin of beer leaving the brewery. So if you are going to brew beer for others to retail (just as per the United States following prohibition) then it is a pretty good idea to be big and powerful and brewing very large volumes. One by one the Big Six fell away, the last being Scottish & Newcastle, who embarked on the bulk brewing route, bought big time into Europe, but finally succumbed to the overtures of Carlsberg and Heineken in 2008, who divided up the spoils between them (thereby making yet bigger global brewers, of course).

Bass? We were already significant hoteliers, with the Crest group. But the board decided that they would really embrace beds and bathrooms rather than barley and bottles and morphed into Intercontinental Hotels, with more rooms than any other company on the planet and with brand names like Crowne Plaza and Holiday Inn. The breweries and brands, as we have seen, were sold to Interbrew before some were divested to Coors.

Huge pub chains were established on the carcasses of the major brewers, companies like Wetherspoons, Walkabout, All Bar One, and the Eerie Pub Company. They make huge investment in the pubs—to the extent that these, while ostensibly more attractive, comfortable, and refined than the hostelries they replaced, have in reality become little more than cheaper family dining haunts. And the beer? Why, the pumps one sees so often are delivering a relatively few global brands.

✿ ✿ ✿

At the last count there were 42 brewing companies in the UK that owed their origins to a time before 1970 and around 600 founded since then. All of the new guys are relatively tiny and, in global terms when compared with the world famous names, so are most of the bigger UK brewing companies also relatively small. Anheuser-Busch/InBev,[8] Molson Coors,[9] Heineken,[10] and Carlsberg[11] are all global entities, with a mighty presence in the UK not least because of Thatcher's legacy. Of the British brewers, the biggest are Marstons (headquartered in Wolverhampton[12]) and Greene King.[13]

It remains possible for brewing companies to own their own pubs—up to a point. And so at the last count Marston's owned 2,203 such hostelries and Greene King possessed 2,091. But compare and contrast with the Independent Pub Companies who, because they do not brew, can own many more pubs—companies like Punch Taverns, owning 8,420 and Enterprise Inns with 7,700.

The beer business is a very different one post-Thatcher, and one tragic victim is cask-conditioned ale (see endnote 19 of Chapter 1). People often ask me what I miss from my homeland. High on the list is this style of beer. Whenever I find myself back in England and whichever city, town, or village I am in, I head straightway to a pub. And if I do not see beer handles,[14] then I turn tail and go in search of them. When Thatcher severed the tied house link, the death knell was sounded for some great beers. Ironically, one clause of the Beer Orders allowed for pub owners to have a "guest" cask ale from another company. However, any self-respecting brewer of such beer demands total control of the product from brewery to glass and would never trust it to a third party.[15] The beer needs to come into condition and also needs to clarify. So the barrels need to be set up properly in the cellar and given time to present to best advantage. And then, once tapped, the contents need to be used within three days; otherwise, we have vinegar, because air-loving organisms called acetic acid bacteria get to work. Small wonder that we at Bass, like others, had an army of "Outside Quality Control" personnel, whose job was to patrol the pubs, making sure that the cask beer was in tip-top condition but also that everything else in the outlet was the best it could possibly be: The glasses were clean; the sparklers[16] were adjusted properly; all beers including those in keg and bottle were also served properly.

I cannot entirely blame The Iron Lady for the demise of cask ale. Some of the guilt must be with technical gurus in some companies, and I will single out Guinness[17] here. For the longest time they had been dispensing their glorious stout under a mixed gas top pressure of carbon dioxide and nitrogen for the simple reason that the latter gas affords much smaller bubbles and vastly more stable foams than does carbon dioxide alone. The research whizzes in Dublin and Park Royal, London[18, 19] set their sights on transferring this mixed gas technology to the canned beer, so that people who wanted to have their draft Guinness in the comfort of their own home could do so with just the same amazing head. The problem is that the much lower carbon dioxide content of draft beers means less inherent foam-forming ability. On tap, gas release can be encouraged by using a pump. But what to do in a can? Enter the widget, the plastic or metal insert in cans that serves as a nucleation device, in plain terms a contraption for making bubbles.

The brewers making ales took note and in no time had produced cask-ale alternatives for the can, pasteurized of course and therefore straightway at variance with the traditional product. Worse, however, is that nitrogen scuppers hoppy aroma for entirely unknown reasons. But there is no disguising it, and so the magnificent dry hop character of ales (something not found on Guinness) is lost.[20] Furthermore the beer develops an extremely smooth mouthfeel.

Undaunted, the product development folks in breweries had their mental light bulbs turned on: Let's pasteurize our draft ales, put them into kegs[21] with less carbon dioxide but also some nitrogen to ensure stable foam and a real s-m-o-o-t-h texture. Dispense it stony cold, of course. Call it Nitrokeg. None of this finings stuff, and natural conditioning and handfuls of hops in barrels. So much easier to clean out the kegs when they get back to the brewery. No skill needed on the part of the publican. No brainer. Cask ale's death knell was well and truly rung.

Of course, with stouts and ales of the purportedly traditional type now available to take home, who needs a pub? Solitary, nonsocial drinking has never been easier. And, in turn, the power of the supermarkets has never been greater. *This* is the new monopoly, the shift of purchasing away from the pubs (in the case of beer) and from smaller, family-driven shops in the case of groceries, clothing, and most other products. Customers are understandably attracted by one-stop shopping at heavily discounted rates. Those cost savings are facilitated by the megastore buyers squeezing their suppliers (e.g., brewers) to the margins. The bigger the brewing company, the more able they are to compete through the economies to be achieved by brewing on a large scale, thereby allowing a modicum of profit from the transaction.

And so off troop the shoppers back to their central-heated homes, with their 60-inch screen televisions to eat their prepackaged fast food and overly cold canned beer as they watch wall-to-wall soccer (so much cozier and cheaper than visiting the stadiums with their extortionate ticket prices). No dropping into the pub at any time—before or after a game, to socialize, to chill out. In any event, you can't smoke in the pub anymore, whereas you can puff away to your heart's content in the home. Small wonder that British pubs are closing at a rate of six every day. No surprise to see the radical shift from draft beer to that in bottles and cans (see Figure 2.1). Unsurprising too, that the percentage of total UK beer sales in the form of cask ale was less than 6 percent in 2008, whereas it was around 17 percent when Thatcher came to power.

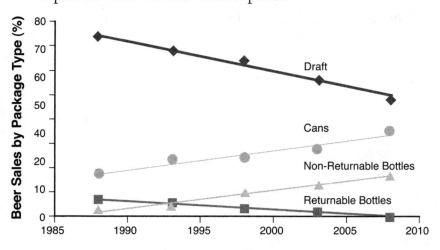

Figure 2.1   Beer trends in the United Kingdom.

You are probably getting the idea that Thatcher is not my all-time favorite politician. The reality is, however, that despite there being a Parliamentary Beer group,[22] there is a long history of the exchequer bleeding the industry dry on excise taxes (duty).[23, 24]

❋ ❋ ❋

The graffiti seems entirely apposite for me: "Nostalgia ain't what it used to be." What right do I have to expect that my notion of the sublime beer-drinking experience is the right one? A rustic pub, cask ale, simple fare—perhaps some fresh-baked ham on crusty well-buttered bread with a pickled onion or two—a log fire, muzak-less buzz of conversation, my *Times*. Perhaps I am growing ornery and am dwelling excessively in the past. Yet I see my glass of traditional ale as a metaphor for spiritual drift. Quiet contemplation, even meditative sipping of a liquid that evolved glacially over centuries swept away by super-chilled beers that seem to speak entirely of this frantic, Blackberried, and iPodded multitasking, less-compassionate world.

3

Barbican, Balls, and Beyond

At Preston Brook we did not brew traditional English cask ales, but some worthy products certainly went into package, notably the nation's biggest selling beer, Carling Black Label, Tennent's Pilsner (fermented with a yeast smuggled from the Czech Republic in a senior manager's handkerchief[1]), and Lamot Pils, a stronger alcohol product.[2] There was the occasional product that I did not care for, however, and top of the list was undoubtedly Barbican.

I am always baffled why anyone would actually want to drink a beer that is genuinely alcohol free (the UK law demands less than 0.05 percent alcohol by volume [ABV] in such products, rather less than arises by spontaneous fermentation in many fruits). Well, actually, I did used to drink it, after playing in cricket games, but only after it had been blended 50:50 with Carling, thereby rendering a reasonably drinkable potion of 2 percent ABV. But there are those who do actually choose such products, ostensibly because they are *seen* to be drinking beer (the peer pressure mentality) or because they have had a problem with alcohol but still want to taste something that is at least from the beer genre, while being free of ethanol. To me it always seems more logical on occasions that I don't want to take in alcohol to actually choose a beverage designed to be devoid of the stuff, like a soda. For the reality is that alcohol-free beer (to my palate at least) is...well, let's just say that such beers make me appreciate how much I like "proper" beer. The simple reality is that alcohol makes a direct contribution to flavor, but also an indirect contribution (it influences the extent to which other flavor-active materials find their way into your nostrils[3]). Take away the alcohol, and there is (within current technology at least) absolutely no way to deliver a product that is truly beerlike (although we could philosophize about what "beerlike" actually means when one considers the enormous diversity of beer styles and flavors in the world). As one of my favorite authors, Simon Brett, has his hysterical actor-sleuth character Charles Paris say in *What Bloody Man Is That*: "No. I may have few principles, but the idea of alcohol-free lager offends one of my deepest. It's like... yuk, I don't know...like the idea of making love to an inflatable woman."

We used to make Barbican by producing a full-strength lager beer and then stripping out the alcohol at low temperatures in a vacuum, under which conditions off would also drift the flavor-active volatiles. The trick was to add back a blend of volatiles, but the absence of alcohol means that no matter how skillful the flavor technologist, the reproduction of the flavor pre-evaporation cannot be achieved.

Above all, I have always been perplexed about how we managed to secure Saudi Arabia as a significant market for Barbican and also for a frankly more tasty variant called Barbican With Lemon, in which added lemon essence successfully disguised the less desirable nuances of aroma in the Barbican base. But Saudi? Overtly Muslim. Why on Allah's earth would they even countenance a product that did not taste particularly good and that was clearly marketed from a beer stance? Was it that the Saudis sought to emulate Western culture while still remaining within the dictates of their faith? Was it that they perceived the product to have a goodness (perhaps they had read about malt and hops and their wholesome attributes) that was not available in their existing portfolio of drinks? Or might they even have liked the flavor? I still do not know, but I am certainly aware that we twice ran into serious problems with the importers over technical issues, one of those occasions demanding my presence in Saudi Arabia. The first problem was when the beer threw the wobbliest of precipitates that looked like frogspawn in the bottom of the bottle. Beer—even the alcohol-free stuff—is not supposed to sit on a dockside in 50 degrees Centigrade, and that is what was causing the problem, rectified by removing foam stabilizer from the product.[4]

Another time we contrived to have some bacteria in the Barbican.[5] I was dispatched to Saudi and must confess to being uncomfortable. It wasn't concern about the bacteria—I could explain that, and they certainly weren't harmful or alcohol forming. My discomfort was all to do with the customs I was encountering. Religious tolerance makes me understand if not appreciate, I guess, the head-to-toe coverage of women. But I confess to outrage when I bought an imported copy of the *Daily Mail* (was there ever a more conservative English paper?) to find that the retailer had gone through the tabloid, page by page, and obliterated all bare female flesh (necks, arms, legs) with a black felt tip pen. It was fascinating on the return flight to watch the dash for the restrooms as soon as the fasten-seat-belts sign was turned off, to see women emerging in tight jeans and high heels. There were other things I didn't enjoy over there—like accidentally kicking over a charcoal burner that was volatilizing fragrant oils, the glowing embers scattering across a beautiful carpet in a tent out in the desert. That's where I had been taken as a guest of honor, to be fed with delicious meatballs that I later discovered had once swung proudly between the rear legs of a great many sheep. Boy, could I have used a beer.

This, then, was one of the less delightful visits I have made in a career that has taken in many of the world's great beer markets.[6] The biggest bazaar for beer is now China. An indication of the astounding growth in the beer business is given by Table 3.1. There are of course a great many Chinese people (at the last count around 1,318,000,000), and cumulatively they do not drink a great deal (some 29.7 liters per head every year). There are good reasons for this—partly financial but partly physiological.[7] Perhaps related to this is a trend ever towards lower and lower gravity (as well as less malt and bitterness) in the beers, as I witnessed in November 2009 when again teaching in Wuxi. And it certainly seems to me that there needs to be a lesson in beer presentation, the notion of cold lager being one of providing the customer with a beer, a glass, and whatever quantity of ice they need to add to chill things down.

One of the most famous Chinese brews of course is Tsingtao, named for a German port on the Chinese coast. Herein was one of the first examples of technology transfer, Germanic brewing principles, in the land of kiu.[8]

American, Australian, and British megabrewers entered into the Chinese market through joint ventures, starting in the seventies—companies like Anheuser-Busch, initially with major shareholdings in the likes of Tsingtao and later through becoming majority holders in companies like Harbin; companies like my own Bass, who bought into a brewery in northern Jilin province. And others.

TABLE 3.1 Changes in Beer Production (Million Hectoliters) in a Selection of Countries

Country	1970	1980	1990	2000	2007
United States	158.0	227.8	238.9	233.5	233.4
United Kingdom	55.1	64.8	61.8	55.3	50.4
China	1.2	6.0	69.2	220.5	393.1
Denmark	7.1	8.2	9.0	7.5	8.0
Germany	103.7	115.9	120.2	110.4	102.2
South Africa	2.5	8.3	22.6	23.9	26.5
Japan	30.0	45.5	66.0	71.7	63.1
Canada	15.8	21.6	22.6	23.1	23.9
Brazil	10.3	29.5	58.0	82.6	103.4
Mexico	14.4	27.3	39.7	57.8	80.9
WORLD	648.1	938.6	1,166.0	1,393.1	1,800.8

The challenges were immense, arising from profound cultural differences. Imagine if you will the logistics from the early days with beer produced using the latest western technology being transported away from the brewery on myriad bicycles. The challenges of acquiring the necessary quantities of suitable grain and hops were immense. However, vast progress is being made, and there is a very real shift in the technical presence in this vast country.[9] Thus we arrive at 2010 to find that a brand, Snow, unheard of outside China, is now the biggest selling beer in the world.[10, 11]

China's growing world rival is, of course, India. Alcohol is a confused entity in the subcontinent where, unlike China, religious fervor and belonging still runs deep. There are real concerns about alcohol abuse, especially by abusive husbands.

The beer industry has lobbied successfully for a proportional reduction in taxation of beer, the drink of moderation, as compared to the dubious distilled spirits that are the root of excessive drinking in India. As a result, the beer market is growing, and as was the case for China, Western countries are investing heavily. Production volumes grew 50 percent between 2003 and 2007; even so, volumes remain modest at 9 million hectoliters, for a population of 1.17 billion people.

As anyone who has eaten in an Indian restaurant in the west knows, it is the norm to accompany the papadums and onion bhajis and chicken tikka masalas with beer served in 750mL bottles and shared between the diners—a very sound notion which speaks well to my passion for beer as a social ritual, as a meeting point for mellowing interchange and conviviality.

I have made many joyous visits to India, flying into the bustling, cosmopolitan, and conflicted Bangalore and journeying thence by train to Mysore, the very act of which transports one back to Imperial times. Such squalor and poverty can be seen through the train windows (if you are fortunate enough to be in a carriage where the glass has been cleaned). The vast majority of people in this country of amazing aromas and traditions cannot afford beer. Growth in the market will perforce be linked to acceleration in the standard of living. It will happen.

What we know as visitors to this spiritual place is that it is perfectly safe to drink the beer, but not the water.[12] Finding a good packaging run was always our first task in the bar: "Bring us beer with these packaging dates only please." Inconsistency was rife, from the heinous diacetyl[13] to slivers of glass in inadequately washed new bottles.[14] For all that, India has come a long way on the beer front.

I wonder: For someone so enchanted by and adherent to the notion of traditional values, with my adherence to cask ale, am I hypocritical in supporting the march of beer into cultures where it was not historically a norm?

The average per capita beer consumption of the Chinese is a little shy of 30 liters per head—and growing (see Table 3.2). In India it is less than 1 liter per head per annum (they drink a lot more whiskey and country liquor[15]). So there is some way to go before either country catches up with the Czechs (see Table 3.2), who consume more beer per head than any other nation worldwide. But a glance at the table reveals worrying signs there and in other great beer-drinking cultures, notably Belgium and Germany.[16] Why? Certainly there are growing pressures over the globe. The UK, Ireland, and Australia among others have fortified considerably the drink-driving legislation. Most people would agree that this is a very, very good thing. Yet more than one Irishman has lamented to me that gone are the days when a farmer in the rural wilds of the emerald isle could trundle down to the remote pub and quaff two or three pints of Guinness before making his way home through the damp night.

TABLE 3.2 Changes in Beer Consumption (Liters per Head) in a Selection of Countries

Country	2000	2003	2005	2007
United States	82.4	81.8	80.6	82.6
United Kingdom	96.8	101.3	95.6	87.7
China	17.7	19.8	23.6	29.7
Denmark	102.2	96.2	90.1	88.5
Germany	125.5	117.7	115.2	111.7
South Africa	53.8	51.1	53.5	56.6
Japan	55.9	50.9	49.6	49.2
Canada	67.4	68.4	67.8	69.7
Brazil	49.6	45.9	49.0	54.4

TABLE 3.2   Changes in Beer Consumption (Liters per Head) in a Selection of Countries (continued)

Country	2000	2003	2005	2007
Mexico	51.0	51.7	55.5	58.4
Belgium & Luxembourg	98.8	96.6	91.0	86.0
Czech Republic	158.9	161.0	164.1	158.8
Ireland	128.0	118.0	106.0	106.0
Russia	37.9	51.4	62.4	82.2
Australia	90.0	87.3	90.9	84.5

Another impact of course is what the drinks analysts Canadean call "Share of Throat." Thus we have wine, which somehow always manages to attract a better and more positive press than does beer. In the majority of countries listed in Table 3.2, wine consumption per head is growing. Even in the Czech Republic and Belgium. Especially in Ireland. What *is* the world coming to?

But it is not only wine. It is other drinks, in this incessant race for fun, fun, fun. Thus the advent of the malternatives, alcopops, flavored alcoholic beverages (FABs)—whatever you want to call them—and we visit them later.

✵ ✵ ✵

Throughout my career I have been privileged to visit with brewers and study or teach brewing in many countries. Wherever I have gone I have discovered my particular joy to be when coming closest to traditional culture. Whether it is a hefeweissen with sausage and pretzel for breakfast in Germany;[17] or stout and oysters in Ireland; or even no beer at all, for instance, cloudberry liqueur in Finland and sake in Japan, even Vinho Verde in Portugal. Of course it is entirely possible for the product of one culture to become assimilated by that of another—witness port passed after dinner in an English black-tie dinner.

And so I am perfectly comfortable with beers brewed according to the traditions of a given nation in a country far from its origin, for it speaks to me of an admittance of excellence. So I delight in Singha beer from Bangkok, brewed to the finest of Germanic and Bohemian lager-brewing traditions and almost miraculously becoming the sublime accompaniment to Thai food. And I admire and savor the achievements of many a brewer of craft ale (and lager) in the United States. Time to come back to these shores.

4

On the Other Hand: The Rebirth of a Beer Ethos

I had never before met Ken Grossman, the owner of Sierra Nevada Brewing Company (see Figure 4.1). It was the spring of 1999, and he invited me to travel the 100-or-so miles from Davis to Chico to come talk about beer freshness with his team. The room was packed with brewers senior and junior, quality assurance personnel, and who-knows-who-else in the impressive bunch that Ken had assembled. And there was Ken, I assumed to take a back seat role as his technical team pumped me for information. But it was Ken who kicked things off with the immortal question, "So, Charlie, what do you think about electron spin resonance spectroscopy?"[1] As he fired question after question at me I was learning very quickly one of the main reasons why his company had grown so phenomenally—and why it produces beer of a quality that is at least the equal of any other brewing company I know.

Figure 4.1   Ken Grossman when awarded an Award of Distinction by the College of Agricultural and Environmental Sciences at UC Davis in 2005. (Left to right: Michael Lewis, Ken Grossman, the author.)

Later that day, I delighted in conversation with Ken over lunch in the first-rate restaurant that adjoins the brewery. He insisted that I try the beer sampler, in which small portions of the many worthy brews of the company are presented to the customer. I tried them all—and recognized each in its own way for its excellence, but I felt obliged to say, "You know, Ken, some of your beers are just about at my upper limit for hoppiness." He calmly looked back at me and, with the glimmer of a smile on that bearded face, replied, "Charlie, 25 years ago I was brewing in a bucket. Now I am producing more than 500,000 barrels every year[2] and selling into every state in the nation. Do you mind if I leave things as they are?"

Sierra Nevada is indeed a phenomenal success story. Ken Grossman was born in Southern California and developed very early interests in matters technical. He was only 11 years old when he built a guest house annex to his parents' home and was not much older when a neighbor interested him in fermentation processes. He repaired electrical and electronic appliances. Moving to Chico in 1971, he combined employment in a bicycle shop with studies at Butte College and California State University, Chico (he never did graduate), and he opened a home brew and winemaking retail store. At college he enrolled in classes with workshops and was able to construct his first brewery (co-owned with Paul Camusi), including drilling all of the holes in the false bottom of his first lauter tun. By day he studied, by night he brewed, and then he would walk the streets of Chico, going from bar to bar endeavoring to interest them in the local ale, with its distinctive grapefruit-like aroma due to the Cascade hop. The Sierra Nevada Brewery was named for his favorite hiking and climbing territory and opened in 1980.

Ken will tell you that it was a struggle, requiring that he and Camusi return several times to friends and family to borrow money to keep the dream alive. He searched dairies as well as breweries the length and breadth of California and even as far away as Germany, to find vessels that he could use or modify as he sought to nudge the production volumes forward. Ken could not afford the carbonator that most breweries employ to bring the finished beer to the specified carbon dioxide level. So he resorted to natural conditioning, the approach that he adheres to to this day, and another special element of the Sierra Nevada package.

There is surely not a feel-good story anywhere in the world that does not feature somewhere a remarkable coincidence, unlikely event, or unexpected turning point. For Sierra Nevada it was when a *San Francisco Examiner* journalist wrote a piece about them in the Sunday color magazine— accompanied by a photograph of Ken and Paul perched on kegs on the front cover—at the time that the beer buyer for the Safeway supermarket chain visited his Chico State-enrolled daughter and fell in love with the Pale Ale. Sierra Nevada was off the blocks and running. Fast.

As volumes grew, so did the wily Grossman progressively increase the sophistication of his facility, in due course moving to a new plot with enormous scope for expansion on East 20th Street. And now, years after buying out his partner, it is the seventh largest brewing company in the United States and truly the most beautiful facility in the world: gleaming copper vessels, hand-painted murals on the brewhouse wall, intricately designed porcelain tiles in the corridors, the very latest in fermenter and packaging capability, a fleet of delivery trucks, a rail link that delivers malt direct to the silos from Canada.

As well as being role models for excellence in brewing, Sierra Nevada is also ahead of the game environmentally. Its commitment to green issues spans from the very simple to the sublime. It stresses the importance of recycling the fundamentals: office paper, cardboard, glass, stretch wrap, plastic strapping, construction materials, pallets, and hop burlap. It seems that in 2006, Sierra Nevada avoided sending 97.8 percent of its total waste to landfill. Sierra Nevada was one of the first regional breweries to employ a vapor condenser to recover energy from the kettle, diverting it to preheat process water. It has effected a modernization of its boiler systems to increase energy efficiency and minimize emissions. It has installed electronic ballast lights and motion sensors, has replaced air compressors with ultra-efficient, speed-controlled drives, and is employing high-efficiency motors and refrigeration systems throughout the facility. Sierra Nevada has cut water usage by approximately 50 percent. It treats all production wastewater to reduce the Biological Oxygen Demand[3] burden on the local municipality, using a sequential anaerobic/aerobic treatment plant.[4] The methane that is generated feeds fuel cells: four 250-kilowatt co-generation power units generating one megawatt of power that satisfies most of the brewery's electrical demand. Co-generation boilers harvest waste heat and produce steam for the system.

The energy efficiency of the installation is double that of grid-supplied power, and air emissions are substantially reduced. The spent grains and trub are fed to cattle that are ultimately transformed into fabulous steaks served in the restaurant adjacent to the brewery. En route, their manure fertilizes Sierra Nevada's experimental hop yard. Small wonder, then, that Sierra Nevada has been celebrated in the Waste Reduction Awards Program of the state of California annually for the past seven years.[5]

Oh, and the beers truly are great. But what would you expect from a man who knows how to do things right—a humble man, who fully acknowledges how the likes of Anheuser-Busch helped him with technical advice as he set out. Don't look in the direction of Chico if you seek to have support for your prejudices against the "big guys."[6]

For inspiration, though, Ken Grossman's role model for the brewing dream was Jack McAuliffe (see Figure 4.2).

Figure 4.2   Jack McAuliffe in the lab at New Albion (note the picture of Albert Einstein on the back wall). With thanks to Sierra Nevada Brewing Company.

In the sixties McAuliffe lived in Dunoon, Scotland, working in the maintenance crew on US submarines. There he fell in love with the local ales and became a home brewer, reading avidly and motor biking extensively in Europe to glean whatever he could about brewing. Upon returning to California in 1968 he went to university before becoming an optical engineer. He remained at his most passionate when pondering matters brewing. Being of an engineering persuasion, he was able to fabricate a brewery in 1975 in the heart of wine country in Sonoma.[7] With a hodge podge of adapted vessels and his mechanical aptitudes sufficient to enable him to construct a simple mill and renovate defunct apparatus, he developed the New Albion Brewing Company, named for the land north of San Francisco Bay described by Sir Francis Drake.[8] The first brews were in 1977, with water trucked down from a nearby hillside. Production was less than ten barrels a week. New Albion lasted but five years, succumbing to the seemingly impossible challenge of carving a niche in an industry dominated by behemoth brewers. Others, like Ken Grossman, overcame the odds. But their inspiration was the big bearded McAuliffe (see Figure 4.3). McAuliffe also inspired the small staff that worked with him. Men like Don Barkley (see Figure 4.4), UC Davis alumnus, who started with New Albion for a wage of a case of beer a week and whatever he felt that he needed to consume while in the brewery. And he learned much, subsequently put to great effect as the brewmaster at the Mendocino Brewing Company, first in Hopland and then Ukiah, with his fabulous Red Tail Ale, and more recently the Napa Smith brewery, rejoicing in the production of a splendid range of beers in the heart of snobby wine country.

Jack McAuliffe and Ken Grossman built their own breweries. Fritz Maytag (see Figure 4.5) pursued a different model.

Figure 4.3   Ken Grossman and Jack McAuliffe sharing memories. With thanks to Sierra Nevada Brewing Company.

Figure 4.4   Don Barkley and the author. With thanks to Diane Bamforth.

Figure 4.5   Fritz Maytag. With thanks to Anchor Brewing Company.

It was the summer of 1965 in San Francisco, and 28-year-old Fritz Maytag, a liberal arts graduate from Stanford and great-grandson of the founder of the famed appliance company, would take lunch at the Old Spaghetti Factory in North Beach. He delighted in drinking a beer on tap called Anchor Steam. Talking one day to the owner of the restaurant, Fred Kuh, he discovered to his chagrin that the brewery was about to close its doors. He hot-footed it down to 8th Street and discovered a facility that was (in Fritz' own words) "the last medieval brewery" and (as he would soon discover) a microbiological nightmare. The company's bank was in the black to the tune of $128.

Naturally, Maytag, who is as much philosopher as he is visionary, fell in love with the place and purchased 51 percent of the business on September 24, 1965. In 1969 he became sole owner, and over a ten-year period breathed new life into the company and its flagship brand.

Fritz Maytag adores history and delighted in the rich tableau that represented the emergence of this San Francisco iconic brewery. Its origins were in the Gold Rush and a German brewer called Gottlieb Brekle. It was the next owners, Ernst F. Baruth and son-in-law Otto Schinkel, Jr., who first used the trade name Anchor. The product was a steam beer, so-called for beers brewed on the west coast using bottom-fermenting lager yeast in the absence of plentiful ice.

The brewery was destroyed in the fire that succeeded the 1906 earthquake, soon after Baruth died suddenly. A year later Schinkel was thrown from a streetcar and killed. Other German brewers Joseph Kraus and August Meyer kept the beer alive before Prohibition closed the business down in 1920. When repeal came, owner Joe Kraus began brewing Anchor Steam again on a new site, at 13th and Harrison, only for another fire to destroy the premises a few months later. Kraus reopened the brewery at 17th and Kansas, just a few blocks from where the brewery is today, and gained a new partner in Joe Allen, who did his best to keep things alive, but bowed to the seeming inevitable in July 1959 and, in the face of the burgeoning might of the nation's megabrewing companies, closed the company down. Enter Lawrence Steese, who bought Anchor and reopened it in 1960 on 8th Street. The struggles continued—until Maytag arrived on the scene.

After steadying the ship, Fritz commenced bottling the brand and launched a series of other beers that are alive and kicking today: Anchor Porter, Liberty Ale, and Old Foghorn. He also produced the first of his celebrated Christmas Ales, with a different and closely guarded recipe every year. And the growth presaged the final shift of scenery in 1977: to an old coffee roastery on Potrero Hill. By 1993, Anchor became the first brewery in the world to have an in-house distillery, making a splendid rye whiskey and a gin.

One of the first things that Fritz Maytag ensured was that he had a laboratory, recognizing that successful brewing is very much dependent upon the application of sound scientific principles. He read Pasteur's *Etudes Sur La Bier*, alongside other seminal brewing texts.[9] He peered relentlessly into his microscope to assure himself that his determination to eradicate the microbiological nightmare that had almost brought the previous regime to rack and ruin was working. Maytag was justly identified as an icon of the brewing revolution in the United States, with his readiness to help and advise others and his passionate exploration of brewing ancient and modern.[10] I confess to a kind of sadness to learn on the very day this manuscript was finished in April 2010 that Fritz Maytag had sold his brewery. The king is dead; long live the king.[11]

Years after the Stanford graduate Maytag had first inspired Californians (and many beyond) to brew beer, a trombone-playing engineering student from the University of California's Berkeley campus who had spent months interning at Anheuser-Busch's Fairfield brewery after a boyhood touring Europe drinking beer (by permission) with his family decided to head to the legendary brewing school at Weihenstephan.[12] Dan Gordon (see Figure 4.6) was the first American to graduate the class on the Freising campus, and he returned to his native San Jose determined to make great lagers according to the Reinheitsgebot traditions.[13] He succeeded, having partnered with restaurateur Dean Biersch ("front of house" to Dan's "back of house" as Dan puts it). The pair opened their first brewpub in Palo Alto in 1988, and the Gordon Biersch legend was born. There are Gordon Biersch brewpubs all over the land—and as far afield as Taiwan. And on East Taylor Street in San Jose lies the main production brewery that delivers a rich array of bottled lagers of every type, year-round and seasonal brands alike, to outlets far and wide. Dan Gordon is a larger than life character in every way. 'Twas he who coined the phrase "never trust a skinny brewer," delivered with his infectious laugh. And it was also Dan Gordon who invented that culinary mainstay of the nation's sporting concessions, the garlic fry.[14]

Figure 4.6   Dan Gordon. With thanks to Gordon Biersch.

Ken Grossman and Fritz Maytag advanced the model of the finest ale brewing, primarily for distribution nationwide. Dan Gordon's model was lagers (and some Germanic ales; for example, his splendid hefeweissen), either brewed in the main brewery for distribution or locally in restaurants bearing the Gordon Biersch logo. Jim Koch, on the other hand, started in Boston, Massachusetts with a franchise brewing approach, underpinning an appeal to US drinkers to be loyal to the flag with their beer selections (see Figure 4.7).

Figure 4.7   Jim Koch in 1986. With thanks to Boston Beer Company.

Koch, a native of Cincinnati and from a brewing family, studied in the law and business schools at Harvard. Struck by the growing sales of imported beers, Koch invested $100,000 of his own money and $140,000 from friends and family in establishing the Boston Beer Company in April 1984. He was 33 years of age. He dug out a time-honored recipe from his great-great-grandfather, Louis Koch, who had been a brewer in St. Louis, a brew adhering strictly to the Reinheitsgebot, and he contracted with the Pittsburgh Brewing Co. to produce it with the name Samuel Adams Boston Lager. He bought a truck and distributed the beer himself in Boston.

In 1987 the beer started to be sold in Manhattan, and a year later he used $3 million from an investment bank to buy an old Boston brewery, though he was stymied by the challenging costs demanded to fill the location with the necessary equipment. It became a tourist location to promote the beers and a center for research and development.

As the 1990s opened, the beer spread into Washington, DC, Chicago, and the eastern states following an agreement for the beer to be produced under franchise by Blitz-Weinhart in Portland, Oregon, a long way from the principal market, although the Samuel Adams Brew House was also opened in Philadelphia.

Koch's marketing strategy played heavily on national loyalty, urging people to drink a fine, fully flavored product from the US rather than imported products. Slogans included "Declare Your Independence from Foreign Beer." They drew attention to the shortcomings with imported beer, namely the fact that beer goes stale if too much time has elapsed between packaging and consumption. The beer scooped a number of awards at the major beer festivals, though his marketing angle irritated not only European breweries importing into the US,[15] but also other brewers in the craft sector. It wasn't long before the big US brewers also started to see the likes of Boston as a major threat.

Production was 714,000 barrels in 1994, and a year later the company went public. By the end of 1995 output had soared to 961,000 barrels, with brands now including Boston Lager, Cream Stout, Honey Porter, Triple Bock, Cranberry Lambic, and a Winter Lager.

In 1997, they purchased the Hudepohl-Schoenling Brewing Co. in Cincinnati, a location that had been brewing its beer under contract, as well as an abandoned brewery in Boston, seeking to dispel criticism that the famed Boston beer was brewed a long way from home. The beer was being brewed in a diversity of locations, including Pittsburgh Brewing Co., F. X. Matt, Genesee, and Blitz-Weinhart. The product range was starting to be exported, and indeed brewed under license at locations such as Whitbread in England.

Small wonder that they now launched Samuel Adams Boston IPA, a beer with a pronounced English provenance.[16] And very soon they joined the light beer momentum with Samuel Adams Light. Nowadays two-thirds of the beer is brewed in Cincinnati.

But what of the home brewer? Here, surely, the doyen is Charlie Papazian (see Figure 4.8). Papazian was a nuclear engineering student at the University of Virginia when he began home brewing. Going to teach in Boulder, Colorado, he continued to brew his beer, spoke about his passion in courses at the Community Free School and wrote a pamphlet to encourage others to take up the hobby. In October 1978 President Carter signed into legislation the freedom for people to brew beer at home,[17] and two months later Papazian founded the American Homebrewers Association, together with the *Zymurgy* newsletter. In spring 1981 he sponsored a brewing conference, which morphed in due course to the Craft Brewers Conference. In May 1982 he launched the Great American Beer Festival, and in 1983 the Institute for Fermentation and Brewing Studies, which would become the Association of Brewers, with its journal the *New Brewer*. These days all of the Papazian operations are within the umbrella of the Brewers Association, with a sizeable staff, a large publishing portfolio, and an ongoing commitment to champion beer, for those operating in their own homes all the way through to the biggest breweries in the craft beer sector.

Figure 4.8   Charlie Papazian. With thanks to Charlie Papazian.

To attend the Craft Brewers Conference or the National Homebrewers Conference is a delight. The atmosphere is humming with bonhomie. The delegates are sponges for both information and beer. The joie de vivre is tangible, the empathy inspirational. Here is a joyous, almost spiritual celebration of malt, hops, yeast, and water. The only disagreeable moments come when someone tries to take a pot shot at the "big guys," who seem to be, for some folks, the devil incarnate. Such attitudes are surely wrong, for brewers in companies of all sizes are united in their determination to produce great beer.

Which leads us to the question: What does quality mean in beer?

5

So What Is Quality?

The table linens were starchy white. The cutlery gleamed. The menus were sumptuous. My hosts and I, in one of the better restaurants in the Netherlands[1], pored over the abundance that was on offer. I made my choice, a straightforward one: a steak.

"And how would you like that done, sir?" asked the somewhat severe waiter.

"Well done," I replied.

He looked blankly at me.

"That is not possible," he said. "Choose something else. Fish, perhaps."

Possibly the English accent was a giveaway. At least he had taken one step back from "fish and chips, perhaps."

Herein lies one of the great realities on this earth, one that this pompous waiter and all too many in our world fail to grasp. Beauty truly is in the eye of the beholder. What is nectar to one palate may truly seem as poison to another.

For my part I am somewhat appalled to see blood seeping from a (to my perception) underdone steak. I savor the robustness of caramelized beef, especially when it has a rich depth of juicy fat at its rim. How I shiver to recall the time a well-meaning Japanese colleague invited me to his Tokyo home to honor me by serving shabu shabu, in which one is presented with strips of raw beef that you are invited to waft for a second or two in a bowl of hot water before deftly chopsticking them into your mouth. How I cringe to remember the time that a Master Brewers Association of the Americas[2] organizing committee in Cincinnati decided that steak tartare was a good idea for a closing banquet.

And yet there are countless people around the world who rejoice in rare or raw meat.[3] Good for them. It is called tolerance. I will allow them their idiosyncrasies. But please, would you do me the courtesy of allowing me mine?

Each to his or her own. And so in the world of food I dismay at the invariable sneer that accompanies the words "British food." It happens all too often here in the United States, a nation that of course gave the world such sophisticated dishes as peanut butter-and-jelly sandwiches and the hamburger. The first of these I detest, the second I adore (well done of course—I cringe when friends order such a minced beef concoction medium rare). Just, in fact, as I adore cockles (see endnote 4 in "Introduction") and Heinz baked beans-on-toast;[4] and Cornish pasties;[5] and Lob Scouse;[6] and Ploughman's Lunches;[7] and tripe[8] (cold with lots of pepper and vinegar); and roast pork after the pork has been dipped in brine;[9] and of course the ultimate British triumph, curry.[10]

Who has the right to say that English food is naff? The French, with triumphs such as frogs' legs, snails, and fungi dug from the ground by pigs? The Germans, with vast lumps of meat, belly-swelling dumplings, and red cabbage?

It is easy, of course, to be sarcastic and demeaning in this way. Indeed, one might say that I am being hoisted by my own petard writing this way. I seek to make only the point that whether it is food or any other manifestation of this earthly existence, the beliefs, passions, and indeed, choices are endless. And who is to say what is good or bad? Let us be mindful here. Let us each respect and be tolerant of each other's opinions.

And surely it should be this way with beer?

What Is a Good Beer?

I (and others) have written at length about beer quality.[11] But what exactly does it mean to the consumer? Most importantly, does anyone have the right to stamp their judgment on what constitutes a good product from a bad one? It is the way in the world of wine with the likes of Robert Parker and Jancis Robinson pontificating on what is and what is not a superlative purchase, thereby at a stroke launching the cost of certain vintages to otherworldly amounts.

As someone I know once said, "Don't you dare tell me, Charlie Bamforth, what a good beer is, just because you have a fancy job title and all those letters after your name. I happen to like my beer with tomato juice and chili and salt."[12]

It works the other way, too. I lose track of the number of times that people have sneered at the famous United States lager brands. The reality is—in my opinion, of course—that these are some of the highest quality beers in the world for their sheer consistency from batch to batch. And you try hiding a defect in relatively gently flavored products. (Again, such would be what you perceive a defect to be, but let's say for the sake of argument it is a flavor that you do not expect to taste in that beer.) It is possible to hide a multitude of sins in a robustly flavored beer of deep roasted character and/or high hoppiness. Not so in a low carb US lager.

Let us journey then from the container inwards and dwell on attitudes towards what is and what is not beer quality.

The Container

I recall sitting in a Chinese restaurant in Sydney, Australia, and asking for a Tsingtao. It duly arrived, happily with a glass. But the beer was in a can. That felt wrong. Even to me, who has studied beer flavor and stability for many years and who knows full well that the beer inside will have been the equal of anything delivered in glass, it seemed cheap and out-of-keeping that a beer selected to accompany my succulent Szechwan should be enveloped in aluminum. Beer with food should come in a bottle—or at the very least drawn into a clean cool glass from a tap—but never in a can. Cans are for fishing trips or football games in front of the television.

Purely psychological, of course. Actually, beer is more stable in a can than it is in a bottle. The seal between the lid and the can body is totally gas proof. Not so for the bottle, where air will creep insidiously between the lip of the glass and the crown cork, the oxygen progressively changing the flavor of the beer in the direction of tomcat pee and wet paper. Okay, perhaps there are those of you who relish nuances of urine and cardboard in your lager. If that is the case, surely you expect those delights to be present in every glass? The problem, then, is that the beer flavor is changing. And so the young beer has one smell, but weeks or months later, it will be quite different. Is that not a quality failure? If such were the case in, say, gasoline, in that it fired up your truck wonderfully well when pumped soon after the tanker had delivered it to the forecourt but a few days later it served only to seize up your engine, then you would be less than completely satisfied.

Provided the inside of the can is properly lacquered and there is no opportunity for metal to contact the beer, then the product will be considerably more stable in a can than in a bottle. And as is the case for bottling, there are big efforts to minimize how much air does reach the beer during filling. Nowadays brewers can buy—at a price—oxygen-scavenging crown corks. And a company like Sierra Nevada has turned back from twist-off caps to pry-off ones, simply because they allow less gas to pass.

Brewers are also well aware of another problem with crown corks, that of scalping. The wonderful hoppy aromas associated with some beers can be progressively lessened by the responsible oils adsorbing onto the coating of the caps. This does not happen in cans.

In 2005 I was interviewed in the *San Francisco Chronicle*. I was asked: "If there are 50 beers on tap, what do you order?"

I answered, "Something out of a bottle."

The interviewer, Sam Whiting, frowned. "Why?"

"Because," I said, "if there are 50 beers on tap I worry that they are not being moved through those pipes as quickly as they should be."

If I go into a bar proudly displaying tap after tap, I invariably ask "which is your best seller?" At least there is then a chance of that brand being relatively fresh.

This presupposes that the bar staff know how to clean their lines. Once I was with my wife in a sports bar in the suburbs of Portland (not my customary type of haunt, but we seemed to have hit a culinary wasteland). I ordered a Bud. I took one sniff of the headless lemon liquid and detected the reprehensible (to my mind) diacetyl. This is the whiff of butterscotch, of popcorn (the smell that keeps me out of movie theaters). Now you need to know that my wife is continually berating me for not complaining in restaurants, but not this day. I called the waiter over and said, "I can tell you that this beer will not have tasted like this when it left the brewery."[13] Chewing, she looked at me coldly. "Then choose something else." My wife was relieved that I did not follow with, "Do you know who I am?"

The simple fact is that vast volumes of deficient beer are served on tap in bars the world over. Hygiene, hygiene, hygiene is the indispensable mantra—otherwise, the beer itself will most definitely be indispensable.

There have been dubious practices since time immemorial with beer on tap. In the day in UK, the standard practice was to put the dodgy beer on tap in pubs close to football grounds on Saturday afternoons. That was the time to unload questionable product into the bellies of men and to a lesser extent women distracted by their soccer passions.

As an 18-year-old trainee barman, I was instructed by George, mine host of The Brown Cow in Helsby,[14] how to deal with the slops that spilled over from the pouring of pints from tap handles. Beneath every tap was a tray, into which the surplus foam and beer was collected as it unavoidably streamed down the side of the glass as a full pint with requisite foamatop was delivered. When the trays filled, we tipped them into buckets, which were progressively collected down in the cellar. George told me that the trick was always to put the contents of the buckets, which of course emerged from barrels of several types of beer, into the barrel of mild ale. That was the darkest and least likely to reveal that it has been adulterated. For the longest time I would never buy a pint of mild anywhere I went.

Speaking of English ales, there is of course the matter of warmth. For such is the popular perception: The English drink warm beer. It is, of course, a myth. Traditional English cask ales are usually dispensed at cellar temperatures— although you still may be fortunate enough to find an old-fashioned hostelry where the barrels are positioned behind the bar. Cellars are below ground, and (in case you did not know it) England is not the warmest country on the planet, and it's pretty darned cool beneath the earth. So the classic serving temperature for cask ale is 12°C to 14° C (53°F to 57°F) (see footnote 49 of Chapter 1).

The Foam

In 1989 I founded the European Brewery Convention Foam Sub-Group, a collection of scientists from across Europe (and North America) who studied bubbles.[15] Sad, you might think. The boffins coming from Germany and Belgium dwelt upon the irony that the chairman was English. For they, like so many non-British, believed that to visit London was to visit the nation. And there, of course, the ale is like tepid tea. Beer of low carbonation is bled into glasses that are exactly a pint from bottom to top and therefore to satisfy the frugal Londoner, there is space only for liquid beer and little (if any) foam.

Such is indeed the norm for cask-conditioned beer in London. But that is certainly not the case farther north. The foam is *de rigueur* in Newcastle and Sheffield and all over the rugged north, both sides of the Pennines. And for the Northerner (especially in Yorkshire) being more money-minded than the "soft" Southerners, then to satisfy the twin demands of a full pint *and* a substantial foam there is a need for oversized glasses, with a line to mark one pint of liquid, but with space atop for the head.

Questions have been asked in Parliament about it,[16] such is the passion of many Britons[17] for a creamy froth. For they are indeed influenced by the appearance of foam.

At Bass we quickly learned this. If we did an in-trade trial comparing two beers and were testing for which had the preferred flavor, then the one that had the best head always came out ahead. People drink with their eyes.

The first thing I did on joining the faculty at UC Davis was an experiment on foam perception. We took cans of a single brand of beer and poured it in a myriad of ways. This illustrates the point that you can create all manner of foam appearances depending on how vigorous the pour is and how rapidly you drain the beer on drinking.[18] More leisurely sipping leads to more foam lacing the side of the glass. Sam Woo photographed the beers immediately after dispense, after half of the glass contents had been siphoned off, and then the empty glass that featured only residual foam. When John Smythe and I showed the photographs in various combinations to people, the overwhelming observation was that beer with good foam scored better.[19] People thought it was better brewed and would taste better. Even folk who confessed to the (to my mind) heinous crime of quaffing their beer straight from the bottle or can were prepared to admit that beer with a foam looked better. The only aspect of the foam that fetched disagreement was the cling. Whereas most men liked to see foam sticking to the side of the glass, many women didn't.[20] They thought it looked dirty and was a sign of a grubby glass. In fact the opposite is true.

There is a sidebar to these experiments. Anheuser-Busch, who endow my professorship, had invited me to St. Louis because they were donating me a slew of instrumentation to stock the dark-ages laboratory that I had been bequeathed in

Davis.[21] While in St. Louis I was invited to make a presentation, and so proudly showed my photographs. Top man (in every way), the very excellent Doug Muhleman, fired the first question. "That's very interesting, Charlie," he said in his deep Southern Californian tones, "why d'ya use Miller beer?" I had not revealed the identity of the lager used, but it was indeed Miller Genuine Draft. Miller pioneered the use of modified hop preparations that protect its beer against skunking, which otherwise is inevitable in the clear glass bottles that it champions. A side impact of these hop preparations is that they give really stable foams, to the extent of being too coarse to the eyes and minds of some.

Of course, foaming can be overdone. Which of us has not opened a bottle or can of beer and recoiled at the sight of the contents spontaneously spewing forth, drenching our pants and sorely trying our patience? One explanation is that at some point in the recent past the container was dropped or, worse still, deliberately shaken by a so-called friend with a bizarre sense of humor. Relax, breathe. They are all God's children.

Even more insidious is over-foaming caused by materials in the beer that potentiate this "gushing." These can be of various types, but most commonly it is the fault of a small protein produced by a mold called *Fusarium* that contaminates grain under adverse growth and climatic conditions. Fastidious maltsters and brewers will never use contaminated grain.

The Clarity

It was my first visit to Australia and I was in a bar in Brisbane. I chose a Cooper's Sparkling Ale (see Figure 5.1). The cheery chap behind the bar pushed the beer across the bar at me, and

I lifted it to eye level with a frown. "Excuse me," I said, "but I think there is a problem. This beer is cloudy." His smile disappeared, "It's supposed to look like that." And under his breath he added "pommy bastard."[22]

Figure 5.1   Cooper's Sparkling Ale. With thanks to Tim Cooper.

A solid reminder: Not all beers are "bright." Most are, and in these any sign of turbidity is often construed as something growing in the beer, whereas in reality it is far more often due to something nonliving dropping out of solution. By no means, though, are all beers clear. Most notable of course are the hefeweissens, splendid beers best served over a breakfast of white sausage and pretzel in Germany. Hefeweissens are turbid because they retain their yeast (hefe). Some such beers are filtered and are therefore "bright." These are the krystallweizens.

The key for turbidity, just as for all other elements of beer quality, is surely an appearance consistent with what you expect. If the beer is turbid, then for goodness sake, ensure that it is always similarly turbid. If a beer is intended to be crystal clear, then ensure that such is always the case.

Over time, beer will develop cloudiness and perhaps sediments due to a number of causes, most frequently reactions between certain types of proteins from the grain reacting with tannic materials from the grain and hops.[23] This will be exaggerated by storing the beer too warm and too cold (never freeze beer). Brewers have got very good at ensuring that they remove these sensitive materials during processing, and so clarity problems are few and far between in this day and age.

Perhaps the most poignant example of what a clarity problem can mean was with Schlitz. Joseph Schlitz Brewing Company was founded in 1849 in Milwaukee. Schlitz was "the beer that made Milwaukee famous." It was the second biggest brewing company in the United States in 1976, at a time when it started to spread the word in the corridors of power that it was going to use its technical innovation superiority to surge ahead of the opposition. Invest in us! It started to accelerate fermentations, but the big bugbear was an unrelated clarity problem arising from the interaction of two process aids added in inappropriate juxtaposition. Put more simply, the Schlitz got "bits": white flakes that made the beer look like a child's snowglobe that you tip over and watch the snowflakes fall. There were other problems around the brewery, not least a crippling strike, but the perception was that here was a company that had little concern for its customers. The brewery was sold to Stroh in 1982, the latter being swallowed by Pabst in 1999. In recent years the Schlitz brand has reappeared; retro rules, as we have seen.

The Color

The color of beer is primarily determined by the amount of darker malts employed in the brewing grist. The more intensely malt is dried (see Appendix A, "The Basics of Malting and Brewing") the more color (and intense flavor) it develops. Color also arises through the oxidation of tannic materials in the grist—the same chemistry as occurs during the browning of a sliced apple. Therefore by controlling the grist and by controlling how much air gets into the brew, brewers can control color.

There is a saying in my native Lancashire: "Once every Preston Guild." The Preston Guild is an infrequent civic occasion, and the saying points to its less than regular occurrence. One time, however, it coincided with my stint with Bass, and we had a display there. One of my team, Kim Butcher, for reasons that elude me at this distance in time, decided to test out a hypothesis that folks not only judge beer flavor on its foam and clarity, but also on its color. Kim put some flavorless food coloring into Carling Black Label (the biggest selling brand of beer in UK) and made it look like an ale. When we tasted the unadulterated lager and the color-adjusted beer "blind" there was no significant flavor difference. But when folks in Preston tasted the two beers they ranked them very differently. The beer with the extra color was deemed more "ale-like," with notes such as astringent, burnt, bitter, malty, diacetyl, full, and heavy singled out. What better testimony of the need to control color. Conversely, what more fascinating a reminder that one could deliver brand differentiation downstream simply by adding color. This can be achieved by adding caramel to the beer, of course, but most brewers would prefer to use water extracts of roasted grain that are fractionated to separate the color from the flavor. Using these extracts a beer can be "colored up" without changing the flavor (and vice versa).

The Flavor

I am always baffled when people say to me, "I don't like beer," for they might just as easily say "I don't like food." For the reality is that there is a seemingly endless range of beer flavors and styles.

I have concluded on being greeted after my public talks with this negative view of beer, that the only people who will admit to it are women (perhaps they are more honest) and that for many of them the problem is gas.[24] I am of course speaking of the carbonation of beer, rather than any metabolic consequence pursuant to the consumption of ales and lagers. I guess I can see this point of view, assuming of course that the selfsame people equally disdain Coke, Pepsi, and Seven-Up, which are more highly carbonated than any beer. And the answer of course is to select a low-carbonation beer—such as a draft ale or stout in a can. Or, to pour with vigor and swirl.

Even then, so many of these beer naysayers will say, "Oh, it's more than just the bubbles. I don't like the taste." I can see why a roasty stout or a double-hopped IPA might be a challenge (they are for me sometimes). But a low-carb North American light lager? That's hardly going to offend the senses. Some would say the taste trends towards zero. That I feel would be a tad extreme, but certainly these products do not overwhelm the senses. Having said which, take a look at the biggest selling beers in the world: None of them has extremes of flavor. They reach the common denominator, which is the great majority of people who do not like to have their taste buds and nostrils blitzed by beer or any other component of the diet.

In Appendix B, "Types of Beer," I journey through the enormous diversity of beer styles. Rather more than the red, white, and pink of wines, would you not agree?

6

Despite the Odds: Anti-Alcohol Forces

That Sunday the clergy at St. Martin's Episcopal Church were away on vestry retreat. Services were limited to Morning Prayer, ably read by the somewhat stony-faced but certainly cerebral Bob C. And we had the bonus of a guest preacher. At least I assumed it would be a plus—but then he started speaking. We had a Bible-thumper amidst us, a man with flashing eyes who railed against the sins of the world, vices to his mind conveniently poured into a bottle of booze. So it was piss-and-vinegar and strictly alcohol-free. For 15 minutes he spat out his message about the route that demon drink will take you in a headlong descent to the devil. When the tirade ceased, Bob stepped forward with the parish notices. With no hint of a smirk nor any change of customary tone he announced, "Make sure you don't forget to buy your tickets for next week's marvelous fundraiser in which not one but two professors of brewing will delight us by telling us about the wonderful world of beer."

There is a long history of religious fervor targeting alcohol in the US and beyond. I had a first glimpse of it a very few years ago when I was a speaker in the charming town of Sonora at the annual dinner of the Tuolumne County Farm Bureau. We had dined well, although I did pass on the rabbit stew. My talk had gone well, with gratifying guffaws. When it came to question time, though, an old-timer in a big hat rose slowly to his feet and drawled.

"I've got two questions for you, sir. The first is: Why did our sweet Lord Jesus Christ turn water into wine and not beer?"

I looked back at him. "Because making beer was so much more technologically challenging than making wine."

The old fellow blinked. "I don't think I'll ask the second question."

People, especially wine-bibbers, often tell me that there is no mention of the word *beer* in the Bible. Not so, at least according to newer translations. And thus while the King James Version[1] has 1 Samuel 1:15 reading

> *And Hannah answered and said, No, my lord, I am a woman of a sorrowful spirit: I have drunk neither wine nor strong drink, but have poured out my soul before the LORD.*

the New International Version[2] declares

> *"Not so, my lord," Hannah replied, "I am a woman who is deeply troubled. I have not been drinking wine or beer; I was pouring out my soul to the LORD."*

And so for Proverbs 20:1, in the King James Version it is "Wine is a mocker, strong drink is raging: and whosoever is deceived thereby is not wise"; whereas the enlightened New

International Version from 1984 offers "Wine is a mocker and beer a brawler; whoever is led astray by them is not wise."

We will not dwell on this slight on beer, merely being grateful that it is recognized that, of course, beer was very much around in biblical times.

And so Proverbs 31:6 in the new parlance declares "Give beer to those who are perishing, wine to those who are in anguish." In other words, take wine for a short, sharp counter to misery, but beer is the thing for those who need saving, presumably on account of its superior nutritional status—see Chapter 8, "Looks Good, Tastes Good, and...."

The confusion has been the Hebrew word *shekar* (shequar) that is customarily translated into "strong drink." But the Babylonians made an alcoholic drink called *shikaru*, and interpretation of ancient clay tablets certainly indicates that this was beer.

Of course it was: The first brews were made by the Sumerians 6,000 to 8,000 years ago. So beer had been around for twice as many years before the time of Jesus as have elapsed since his death.

Such debates, however, are perhaps inconsequential when compared with the venom, bile, and extremes of interpretation that flow from the biblical interpretations of the religious right. There is a distinct correlation between the extent of religious condemnation of alcohol and the tendency of a denomination to adopt literal acceptance of the words in the Bible.

Consider, then, the moderate stance of the Rev. Dean Taylor, from St. Mark's Episcopal Church in Georgia:

> *The Episcopal church believes there is nothing innately wrong with alcohol consumption. But we do recognize that the disease of alcoholism can destroy lives, so our ministry to those folks with that disease is pretty important to us....*

Alcohol is part of God's creation and as such it is not
evil. It is only what we do with it. Over-consuming to
the point that we hurt ourselves or other people—
that's when it becomes an evil....

Compare and contrast with the guy down the road, Bill
Walker, pastor of Dug Gap Baptist Church, who opines: "We
are at a loss to find any good that alcohol as a beverage does in
our society. I haven't heard anyone say that it does any
good...."

Taylor must be oblivious to the wealth of news stories talking about the potential health benefits of moderate alcohol
consumption. And can Pastor Bill with good heart condemn
the good Trappist[3] brethren, making their legendary brews,
the last remnant of the ages-old tradition of brewing in the
Church?

But what of other denominations? In the Jewish tradition
there is generally a tolerance of alcohol; indeed, it might even
be judged as a good deed, a mitzvah when consumed in moderation on the Sabbath.[4]

In the third Abrahamic religious strand, Islam, alcohol is,
of course, forbidden.[5]

Quran 5:90:

O you who believe, intoxicants, and gambling, and the
altars of idols, and the games of chance are abomina-
tions of the devil; you shall avoid them, that you may
succeed.

There are actually some more moderate words in the book:

Quran 2:219:

They ask you concerning intoxicants and gambling. Say to them, in them are great harms and only some benefits for humankind. But the harm of them is much greater than their usefulness.

And there is certainly some hope for the afterlife:

Quran 47:15:

The allegory of Paradise that is promised for the right-eous is this: it has rivers of unpolluted water, and rivers of fresh milk, and rivers of wine—delicious for the drinkers—and rivers of strained honey. They have all kinds of fruits therein, and forgiveness from their Lord. (Are they better) or those who abide forever in the hell-fire, and drink hellish water that tears up their intestines?

Okay, not beer, but we live in hope.

In Hinduism, a spiritual tradition based on an understanding of dharma (the natural laws underpinning the universe), there is no ban on alcohol. Typical of the religion, a position is taken of observance and understanding of the merits and demerits. The medical aspect of the dharma, the Ayurveda, recognizes the value of alcoholic beverages when used in moderation.

If only all people of supposed holiness might consider the Hindu way.[6] The approach here is to dwell upon matters case by case—as Frawley says, it is not possible to bring the dharma down to a handful of do's and don'ts. Might I suggest that water is a wonderful thing when it slakes the thirst of a desert traveller, but a force for disaster for the drowning nonswimmer? Likewise beer as part of convivial intercourse is a very different entity from beer within a framework of drunken debauchery.

Despite my Episcopalian pursuits, I aspire to act with Buddhist tendencies.[7] And so it is with a genuine inner struggle that I ponder the fifth "fold" of the Eightfold Path,[8] the pursuit of Right Livelihood.[9]

In all sincerity I have dwelt upon these teachings, but then consider the reality of day-to-day life in Tibet (and other traditionally Buddhist countries in the Himalayan regions) and a widely used product called *chhaang*.[10] How is it made? Barley (or millet or rice) is allowed to ferment in a bamboo barrel before boiling water is poured through it. After cooling, yeast is added, and it is left to stand for two or three days to produce something inappropriately known as *glum*. A little water is added to it—then enjoy!

Hey, that's beer! It is said that the Abominable Snowman likes a drop of chhaang. What better endorsement could one possibly have?

I rather think I rejoice in Buddhist thinking because of its fundamental tenet of tolerance. I also feel that the greater Episcopal Church in the United States also speaks with a mindful and compassionate voice, welcoming with loving-kindness into its fold women priests and homosexuals. And what other church could use the likes of me and my predecessor to raise much-needed dollars by holding beer lectures and tastings?[11]

The word is *mindfulness*.[12, 13] If only there was a more mindful and contemplative approach to life and a genuine acceptance of the greater good in the hearts of people, then surely we would arrive at a gentler and more tolerant society? Let me illustrate.

My student, Troy Casey, and I recently wrote a paper confirming that beer is perhaps the richest source of silica in the diet.[14] Within a very few days of a press release by the journal, there had been astonishing worldwide coverage in newspapers,

Internet blogs, and on television and radio, ranging from Colombian National Radio to the BBC (they ran it on *Five Live* at 2:30 a.m. UK time). In the majority of this coverage the journalists were responsible and sensible. Consider, though, some of the questions I received from people in the United States:

> *Do your findings suggest that binge drinkers have really good bones?*
>
> *Would beer affect growing bones in a child, like milk does?*
>
> *How does drinking one beer for health affect the brain?*
>
> *As a professor in a university, do you think that this study could have students trying to justify their drinking habits?*

And the inevitable

> *The drinking age is 21, but with health benefits being identified, do you feel the drinking age should be lowered?*

Others drew attention to the fact that my professorship was endowed by Anheuser-Busch and inferred that it would be impossible for the study to be impartial.

Whether their gripes were about scientific integrity (and I certainly resent any suggestion that the study was anything other than unprejudiced) or whether they were alarmed at anyone daring to deliver yet more evidence about the potential merits of drink, I am of course unaware. It is undoubtedly a fact, however, that neo-prohibitionism is alive and kicking in the US, to my mind a fearsome example of intolerance. One presumes that they passionately aspire to restoring once and for all the prohibitionism that reigned in the United States between 1920 and 1933. The forces that led to this drastic stance are meaningful and truly speak to conflicts founded on excess.

As Fleming wrote in 1975,[15] the earlier settlers truly believed that human survival depended on the consumption of alcohol. As Fleming wrote:

> Men and women, old and young, rich and poor, regularly started the day with a morning dram. The drink might be anything from cherry brandy to wine mixed with sugar and water, as long as it contained alcohol. A daily glass of "bitters"[16] was considered essential for warding off disease, clearing the head, and keeping the heart in good working order.

Shoppers could take a drink from the barrels of rum on tap in the establishments that they visited. Laborers broke off midmorning for "bitters," and rum was quaffed by those working in the fields. The booze was heady stuff, not the much less alcoholic beer.

Dr. Benjamin Rush, a signatory to the Declaration of Independence, wrote in 1784 that "ardent spirits" caused among other things obstruction of the liver, jaundice, hoarseness, diabetes, jaundice, gout, epilepsy, madness, and "frequent and disgusting belchings." One response came from a group of Connecticut businessmen who replaced the rum for their employees with hard cider and beer.

The Union Temperance Society was founded in New York in 1808. Its members pledged to "use no rum, gin, whisky, wine or any distilled spirits...except by the advice of a physician, or in case of actual disease, also excepting wine at public dinners." No mention of beer—as was the case for several other such societies that were formed, the argument being for moderation rather than abstinence. President Thomas Jefferson wrote to a friend in 1815 about beer: "I wish to see this beverage become common instead of the whiskey which kills one third of our citizens and ruins their families."[17]

It was a Presbyterian minister, the Reverend Lyman Beecher,[18] whose pulpit-thumping inspired the crusade to rid the country of "rum-selling, tippling folk, infidels and ruff-scruff." As a consequence, employers stopped supplying drinks to their workforce, and liquor rations were halted in the US Army. Beecher's American Temperance Union (ATU) strove for a national ban on the sale of alcohol. Up until 1836 beer was accepted within the ATU, but that too was outlawed, and as a consequence the membership declined, the realization being that it was hard spirits that were the real problem, not the drink of moderation, beer.

In 1833 the US Supreme Court ruled that states could regulate the liquor trade within their boundaries. Indeed, individual counties and towns could introduce prohibition if they so wished.

Maine introduced total prohibition in 1851, and before long thirteen other states had joined them, although nine soon repealed the laws or declared them unconstitutional. Only Maine, Kansas, and North Dakota stood firm.

The crusade against alcohol was driven in substantial part by women, in the face of ill-treatment at the abusive hands of liquor-soaked men. Among the Women's Christian Temperance Union's membership was the First Lady, Mrs. Rutherford B. Hayes ("Lemonade Lucy"), although there is no implication that the President had a drink problem.

Discussion of prohibition always leads to Carry Nation, who declared "hatchetation" in smashing up illicit taverns in her home state of Kansas and beyond and set off on an enthusiastically received lecture tour in which hatchets could be bought as souvenirs. They do say that no publicity is bad publicity, and soon liquor producers were marketing Carry Nation cocktails and bars were decorated with hatchets and signs that declared "All Nations welcome but Carry."

The reality is that Carry Nation was neither the most organized nor the most credible of crusaders. The Anti-Saloon League, which originated in a Congregational Church in Ohio, had much more low-key and cunning tactics, progressively persuading towns and counties to embrace prohibition. Soon they were successful at the state level. In 1913, they marched on Washington, DC, with a slogan "A Saloon-less Nation in 1920." Several supporters were elected to Congress.

The 65th Congress, convening in March 1917, declared war on Germany following the sinking of the *Lusitania*. Laws were introduced that would support the war effort, including one concerning the production and distribution of food. A clause was inserted that outlawed the production and sale of alcoholic beverages, such that grain could be conserved. There was disagreement from the opponents of prohibition, and there was agreement to let the Senate vote on a separate resolution calling for a prohibition amendment to the Constitution. The Eighteenth Amendment went speedily through Congress, and it was ratified by 36 state legislatures in little more than a year. Only Rhode Island and Connecticut held out. The amendment was officially adopted on January 16, 1919, with national Prohibition being effected one year later.[19]

The impact? In New York before prohibition there were 15,000 bars, whereas after Prohibition there were 32,000 speakeasies. Booze of dubious provenance was shipped in illegally from Canada, Mexico, Cuba, the West Indies, and Europe. Much of the illicitly brewed drink in the States was dangerous. Bootleggers collected $2 billion annually, amounting to some 2 percent of the gross national product.

Organizations such as The Moderation League sprang up seeking to repeal the Volstead Act. In 1930 the American Bar

Association adopted a resolution that called for repeal. They were supported by the Women's Organization for National Prohibition Reform. Those advocating "dryness" were at risk of being perceived as defending the gangster culture.

Many argued that the Great Depression had been brought on by Prohibition and that to repeal the act would be to create jobs and generate much needed taxation.

The 1932 presidential campaign was substantially fought on the alcohol lobby. Herbert Hoover said that Prohibition had been an "experiment noble in purpose," and he promised to address whatever shortcomings there were. Franklin D. Roosevelt went a major step further: "I promise you that from this date on the Eighteenth Amendment is doomed."

Roosevelt was elected, and nine days later he successfully asked Congress to amend the Volstead Act so that the alcohol content of beer could be raised from 0.5 percent to 3.2 percent by weight.[20] As he sat down to supper on March 12, 1933, Roosevelt is quoted as saying, "I think this would be a good time for a beer."[21]

Fast forward to 1980 and a distraught woman, Candy Lightner, founding Mothers against Drunk Driving (MADD), following the tragic death of her daughter hit by a drunk driver in Fair Oaks, California. Within five years she left the organization, saying, "It has become far more neo-prohibitionist than I had ever wanted or envisioned... I didn't start MADD to deal with alcohol. I started MADD to deal with the issue of drunk driving."

Surely there is the nub? All right-minded people would condemn irresponsibility associated with alcohol. Yet carte blanche declarations that alcoholic beverages are entirely without merit is the frightening stance of some rather powerful forces in the US.

MADD is now one of a formidable number of organizations[22] seemingly bent, if not on absolute prohibition, at the very least on portraying alcohol and the people who deign to drink it (no matter with what degree of temperance) in the most negative of ways.

Frequently there seem undertones of radical religious fervor—witness Joe Califano[23] of the Center on Addiction and Substance Abuse (CASA), who in saying that alcohol's "availability is the mother of abuse" declared "for me, establishing and building CASA and committing myself to this battle against substance abuse was doing the Lord's work."

By way of another example, consider the Marin Institute, self-declared "Alcohol industry watchdog." They advocate major change to policies of advertising and taxation, saying inter alia that the former clearly tempt young people into drinking alcohol and the latter is the route to preventing the abuse of alcohol [24].

One would need to be naïve indeed to argue that much of the television advertising for beer is not potentially dubious at best. The advertisements can be extremely funny, but equally they hardly appear to come from a position of maturity. It certainly has been an uphill battle for me endeavoring to spread messages of moderation and that beer ought to be a beneficial, welcome, and wholesome aspect of an adult's lifestyle when I am confronted by imagery of flatulent horses and soccer ball juggling turtles as an aide to selling beer.

Surely it demands the middle way, from a Buddhist doctrine? I for one see nothing wrong with imagery of sweeping prairies of golden barley or hops growing lusciously on the vine, and with people sipping the ensuing beer in warm and cozy environments.

And yet it seems that organizations like the Marin Institute are as extreme in their doctrines as some brewers are with their marketing.

As regards the Marin Institute's advocacy for a taxation solution to the issue of excessive consumption, I truly wonder. Are those who are compulsively attracted to alcohol, whether or not by cause of addictive behavior or alcoholism, truly going to curtail their habits based on a higher tax slapped onto their beverage of choice? And would this not be a disservice to the majority of people who consume alcohol abstemiously and mindfully? Would it not be as illogical as boosting the taxation of major foodstuffs[25] because some people binge eat and rather too many consume excessively and are, not to pussyfoot around the issue, fat?

Again I plead for a mellow meeting place on these contentious issues—literally the middle ground. But what hope is there for a reasonable debate in an arena where the likes of Beer Soaks America declare that drinking alcohol can lead to "riots, violence, and contracting sexually transmitted disease ."

So can stony-cold sober religion.[26]

7

Societal Issues

The British soccer fans are a little bit naughty when they sing "Two World Wars and one World Cup" to the tune of "Old MacDonald Had a Farm" and in reference to the defeat of Germany on the battlefield[1] and in a famous Wembley cup final of 1966. Yet the rivalries continue, if not in military combat but certainly still in Association Football and in the world of beer.

Fundamentally, there are two great brewing traditions, those of Germany and England.[2,3]

The latest battles it seems are being fought in the world of beer strength, with the Americans weighing in for good measure. In fact let's start in Boston, with the Boston Beer Company's succession of extremely alcoholic beers that culminated in Utopias, 24 percent ABV and more than $100 for every one of the 8,000 cute little brew vessel-shaped bottles. Boston's founder Jim Koch is quoted as saying, "We are passionate about brewing quality, one-of-a-kind beer that amazes beer lovers and aficionados alike." The beer is touted for post-prandial sipping.

I recall asking my students when the beer was launched: "Why?" What is the point? Is it a who-can-pee-highest competition or simply a matter of shooting for the *Guinness Book of Records*? Little did I expect what happened thereafter. The Germans joined the pissing contest with Schorschbock 31 from Schorschbräu with 31 percent ABV. Now it is simply not possible to produce these types of alcohol levels by natural fermentation. And so this eisbock is made by partially freezing a doppelbock. When you start to freeze a beer, the first thing to solidify is pure water, leaving the alcohol and the other components behind. So the trick is to produce just the right amount of ice and then drain off the fiery remainder.

Crazy? Then try Tactical Nuclear Penguin, which soon after came out of the Brew Dog company in Scotland in November 2009 at 32 percent ABV and £30 per bottle. "That'll teach the Germans," they must have thought. What's more, it was based on a beer that spoke to a time when the British empire flooded the world's maps in red,[4] namely a 10 percent ABV imperial stout, aged for 16 months in whisky casks before the freezing process.

"Gott in Himmel," they must have said down in Franconia. "Was ist das?" Undaunted, they unveiled Schorschbock 40, which as the name suggests contains 40 percent ABV. Stupid? Then try the name and the nature of what followed from Scotland: Sink The Bismarck at 41 percent ABV.

By the time this book hits the presses I feel certain that the alcohol race will have taken us yet further into realms of ludicrous idiocy.[5] What better weapon could there be for the anti-alcohol lobby?

This is not moderation. This is not even good and sensible beer (in my judgment).

I enjoy a glass of whiskey. And as so many others do, I take it with water. The dilution allows a fuller appreciation of the flavor.[6] Will you be taking water with your Bismarckian folly?

We do not have to drift to the wilder extremes of alcoholic frenzy to find evidence for brewers shooting themselves in the foot. Consider the world of the malternative, otherwise known as alcopops, or ready-to-drinks (RTDs), or flavored alcoholic beverages (FABs).

In Bass's research department in Burton-on-Trent we had a substantial new product development effort, part of which was devoted to producing beverages based on a bland alcohol base that was made in the brewery in Glasgow.[7] Several products emerged into the marketplace, the first of them being Coral, a product with a lurid color and an overwhelming aroma of passion fruit, a drink marketed with the aid of inflatable parrots. Rather more subtle was Gibson's, which (if memory serves me correctly) had a tropical cucumber aroma and was not only served in slender cans but also on tap in Blackpool, where (legend has it) Scottish holidaymakers knocked it back in pint-loads. We even brought out one with a taste of strawberries, topped with a creamy white foam produced from egg white.[8] None of these drinks lasted long in the marketplace.

I recall going down one time to the London offices of our spirits subsidiary Hedges and Butler and serving up a row of 20 or more bottles of different flavored hopefuls to be launched as the next great alcopop success story. One by one my host sipped the drinks and then he pointed to one of them. "That is the best," he insisted. My jaw sagged. "But that is orange?" I spat, disgusted that the likes of melon-and-ginger, "pina colada," and guava-and-lychee had been passed over. He nodded sagely and uttered a profound wisdom that for some unfathomable reason has stuck with me ever since: "Why sell brown shoes when black ones sell perfectly well?"

In the end one flavor did triumph—and it was of course lemon, the primary product in the Hooper's Hooch portfolio.

The very name has all the wrong connotations for me. And certainly its provenance sat uncomfortably with not only me but also the wiser sages in the company hierarchy. It would be very hard for anyone to argue that these products—and the many worldwide that have followed on from this one and others like Two Dogs[9] launched around the same time—are anything other than drinks focused at the younger element. That would be younger as in "above legal drinking age," of course, but nonetheless they could hardly be said to be positioned at hoary old geezers slowly sipping pints of mild, nibbling on pork scratchings, and reading the *Daily Mirror*.

These products are generally of comparable alcohol content to mainstream beers. As such they are drinks of moderation. However many brewers' concerns are that they are not only understandably interpreted as being party drinks for the immature but also their very existence jeopardizes the long-term existence of beers as they know and love them for the simple reason that younger drinkers who are challenged by bitter products will not gravitate from these sweeter and easier-to-drink alcoholic products. It might of course be argued that traditional beers have no God-given right to always be the "go-to" drink for the masses.

It is small wonder that anti-alcohol lobbyists in the US hit upon these products as a disingenuous way by which brewers expand their markets at the younger end of the age range. And to provide them with famous spirit names while at the same time justifying the taxation rates of beer on the basis that they are brewed products that genuinely accord with all extant legislation, sticks in the craw of many legislators, who would seek to have them taxed at the higher spirits rate.

Simple fruit-flavored alcoholic drinks are one thing—but what risk are brewers running with products incorporating the likes of caffeine?

It may fairly be argued that it is perfectly possible for those of legal drinking age to head to bars and knock back all manner of cocktails. Such is true—but these are not beers and neither have they any genuine heritage in breweries. As we saw in Chapter 6, "Despite the Odds: Anti-Alcohol Forces," the primary prohibitionist forces were spawned in a nation pickled in distilled hard liquor. Beer is customarily a drink of moderation and to my mind remains a product that should be free from ludicrous displays of gung-ho excess such as outrageous alcohol content and foolish ingredients.

Alas, despite the reality that the majority of beers are truly products of genuine moderation, all too often beer and beer-drinking are portrayed in an unfavorable light in many circumstances.[10]

Now lest the reader start to wonder that I am some kind of bigot intolerant of excess, I will confess to more than the occasional foolish act of revelry in my lifetime, not least the time I was poured onto a train after leaving Oktoberfest, thereafter recalling nothing of my journey to Freising (where I was spending time at Weihenstephan) though waking with not inconsiderable horror to find that I was using some disgusted fellow passenger (male) as a pillow.

The Oktoberfest is but one example of a drinking ritual that can lend itself to abuse. Yet, with its good spirit and generally wholesome intent, it is surely more acceptable than unforgivable and unendurable drinking rituals that too often besmirch the good name of beer in places like the United States.

I am talking drinking games. Let us consider a few.

Perhaps the best known is beer pong, for which you need a table, some glasses of beer, and a table tennis ball. Two teams take their places on opposite sides of the table and aim to pitch the ball into the glasses belonging to the opposing side that are arranged like bowling pins. If the ball enters a glass then the contents have to be drunk by the side that has that glass. The remaining glasses are arranged to be close together. Once a side's glasses are empty, then they are deemed the losers and get to deliberate while finishing off the beer of the winning side.

You think that is idiotic? Then try Ten Minute Warning: Every minute for ten minutes you drink some beer. The first minute you take one sip, the second minute you take two sips, the third minute three sips, and so on. You lose if you don't quaff the requisite amount of beer in any given minute. It amounts to rather a lot of beer in ten minutes—and it is stupid. They are but two games—I can't begin to mention some of the others, with names such as Asshole, Blow Me, and Scooby & Shaggy Kegger.[11]

Naturally all of these games are not necessarily restricted to beer; indeed probably they pass with even more debauched hilarity when sips of lager are swapped for shots of hard liquor. And whereas the results with beer are customarily some fairly offensive multicolored vomiting, the terrifying reality when spirits are the playing piece is that the outcome can be a mortuary slab.

Strange. Wine is never the vehicle of choice.

And neither should it be beer.

I say again, I have been no angel, though never once descended into the murky world of pub games. If we are in the mood for confessions, then here goes.

It was 1969, and a barman's eyes showed concern but his lips revealed the faintest of smiles.

"Are you sure you can handle this, lad?"

I affirmed with a disturbing slur in my voice that I most definitely could, and thus was granted my fifth (or so) pint of cider.[12] I shuffled off and sat back to listen to the music, enviously watching as the other sixth-formers laughed and chatted, some dancing modestly. My heart ached to be among them. Instead I had another glass of yellow solace with glazed eyes fronting inner melancholy.

Looking back these 40 years I cannot recall how it could be that a table was set up at a grammar school[13] event, freely dispensing alcoholic beverages to those who were a year or more short of their 18th birthdays.[14] Neither can I remember much else about the evening, until the time I reached my home after a trudge of a mile or more through a driving blizzard. For it was when I got to the front gate that my sins came out, literally, into the snow.

I recall waking up next morning with a pneumatic drill in my head, but rational enough to think, "Oh, shit, what is my mother going to say when she goes out front?" I staggered out of bed, tore back the curtains, and focused (eventually) on the great wide world. It was pouring with rain, wonderful rain, and the snow and all the crimes within it had been washed clean away.

I wish I could say that I learned my lesson there and then (well, actually, I did as regards hard cider and have not touched a drop of the stuff since). But alcohol, in general? I fear not. On a few too many occasions through my life I have gone one over the top with beer or spirits, though rarely wine. Not once did I wake to feel anything other than shame, remorse, and regret, vowing "never again." As I have grown older and very much wiser I have learned restraint and respect for the perils of abusing alcohol, and few and far between are episodes of excess. Equally, I can look back to vastly more occasions lubricated with alcohol in moderation, times that have been suffused with joy, contentment, and gladness.

My cider-drinking exploit was, pure and simple, part of the escape route from the awkwardness, insecurity, and confusion of a fatherless adolescence.[15] The solution should not have been (and is not) the drowning of sorrows. I was after all below the legal drinking age, but more importantly it was an unpleasant and demeaning spectacle. So much better to have sought alternative strategies for assuaging distress and sorrow and hormonal confusions with (at the appropriate time) alcohol being a moderated adjunct as but one of the many pieces that comprise the complexity of life.

All alcoholic beverages, including beer, can accompany occasions of great joy, hilarity, mirth, bonhomie, and joie de vivre. Alcohol is a catalyst towards conviviality.

And I will not be convinced that drinking *causes* aberrant behavior, such as hooliganism, violence, and the like. For sure, it may act in a disinhibitory capacity, but it is not alcohol that creates spouse-beaters or overly aggressive individuals. Such tendencies are inherent in a person's makeup.

For me, in the 30+ years I have been attending brewing meetings around the globe, I have seen (and participated in) the consumption of rather large quantities of beer. I have yet to see a single argument or bout of fisticuffs at any one of these innumerable gatherings. And rare indeed have been sightings of hard-core drunkenness. But I have seen a plethora of entente cordial, from men and women of countless nations.

8

Looks Good, Tastes Good, and...

"So if we could only boost the level of thiamine, would you agree that beer is a fairly well-balanced foodstuff?"[1]

Even for me that would be a fairly tall claim, but a renowned medic was sincerely suggesting it, having heard me discuss the composition of beer at a meeting dedicated to exploring the "Science Behind Health Benefits of Moderate Beer Consumption," an event organized by the University of Maryland Center for Food, Nutrition, and Agriculture Policy in Washington, DC, in October 2006.

I think I replied to the good doctor that at least one would need to accompany the beer with a few pretzels. Tongue-in-cheek stuff, of course.

Count me, then, among the majority of people who take a beer for the simple pleasure quotient. Few people surely make conscious decisions about taking a beer (or any other alcoholic drink for that matter) on overtly nutritional or medicinal grounds. As my student Christine Wright[2] showed, in decreasing order of importance in the decision-making process regarding purchase of alcoholic drinks we have

Taste
Activity—what I am doing
Location
Time of day
Price
Company—who is with me
Food pairing
After-feelings
Alcohol content
Serving size
Weather
Calories
Healthfulness
Carbohydrates

However, it is important that any drinker be aware of the relative positive and negative consequences of their choice. Too often people are prepared to accept that alcoholic beverages are thoroughly bad: a portal to doom, debauchery, and damnation, messages perpetrated by organizations of extreme religious or self-proclaimed moralistic fervor. On the other hand, there are those who ignore any cautionary words of the risks of intemperance.

Edgar Vincent D'Abernon (1857-1941), banker, erstwhile Conservative Member of Parliament for Exeter, director of the Lawn Tennis Association, and the first British Ambassador to Berlin after the Great War, chaired in 1918 the advisory committee on alcohol appointed by the British Medical Research Council, the aim of which was "to separate what is knowledge from what is surmise, conjecture, or popular belief, and, by this preliminary clarifying of the question, to prepare the way for further research." In the preface to the resultant book titled *Alcohol: Its Action on the Human Organism*, Lord D'Abernon wrote

> Alcohol is an ungrateful subject. Most people who are interested in the subject are already partisans on the one side or the other, and no body of impartial opinion exists which is ready to be guided by scientific inquiry. The majority of those who would give any attention to original work on the subject would do so less to gain knowledge than to find arms and argument to support their preconceived opinion.

In 1957, Chauncey D. Leake, professor of medicine at Ohio State University, said at a symposium titled "Alcoholism" held under the auspices of the American Association for the Advancement of Science

> As better knowledge and understanding of the action of alcohol becomes available, more sensible attitudes regarding it are arising, [but] it is also interesting to observe how little the people wanted to learn about alcohol in a scientific way. They seemed much to prefer their violently differing emotional fantasies about it."

For better or worse, the present author has been a member of the brewing industry since June 1978,[3] either in a research, quality assurance, or academic role, and will undoubtedly therefore be condemned by the anti-alcohol lobby as partial and on the dollar of those who they would perceive as cynically pouring alcohol down the throats of the naïve and foolhardy. To my mind, such accusations would be no more nor less astonishing than to argue that a male judge would be incapable of recognizing crimes of a man against a woman (or indeed vice versa).

Lest the reader starts to believe that the author doth protest too much, let us consider with all due impartiality the current state of affairs with regard to beer and health. And let us start with the medical profession, trusting they are at arm's length from funding drawn from the drinks business or (conversely) the anti-alcohol lobby, for the latter could be equally perceived as self-serving.

It is important to stress that the bulk of the conclusions concerning the impact of alcohol on health, whether positive or negative, are based on statistical correlations derived from the questioning of patients or at the very least from observations that are not free from confounding influences. As such they demand close scrutiny and critical appraisal before being set in the gospel tablets of either the pro- or anti-alcohol camp. So much of the data is generated from questionnaires, of the type: "how much alcohol do you drink per day?" The inherent tendency of patients is to underestimate their consumption, a phenomenon more readily described as "lying."

Another highly pertinent problem is that of "secondary correlations." Let's look at a hypothetical example that has nothing to do with beer.

Think of sumo wrestlers. They are all very big blokes. The vast majority are Japanese. So there is a strong correlation between the total number of sumo wrestlers and the proportion of the population that are Japanese. There is also a stronger correlation, I would suggest, between the proportion of Japanese people and the tendency to worship at a Shinto shrine than there is for the general world population. However, this does not mean that to worship at a Shinto shrine will lead to your chances of developing a body fitting you for the sumo dohyo.[4] Writ simply, the uninformed conclusion would be "pray at a Shinto temple and get fat."

Taking a direct example, the World Health Organization reported various statistics purported to link alcohol consumption to violence (http://www.euro.who.int/Document/E87347. pdf). For example, 53 percent of the victims of assault in Norway claimed that their attacker had consumed alcohol before the crime. The data isn't reported, but I daresay approaching 100 percent of the criminals had eaten food before the assault. The naïve would infer that to eat is to predispose one to aggression. Stupid? Of course. But so too, surely, is the assertion that alcohol per se causes violence.

When it comes to drawing correlations between health effects and the consumption of alcohol, therefore, we need to be careful. The benefit or the adverse impact may have nothing whatever to do with drinking. Rather it may have everything to do with other activities undertaken by the drinker—as we shall see.

Doyens in the field of alcohol research include Dr. Arthur Klatsky of the Kaiser Permanente Medical Center in Oakland, California; Dr. Norman Kaplan, Professor of Internal Medicine at the University of Texas Southwestern Medical Center in Dallas; and Dr Curtis Ellison, Professor of Medicine and Public Health at Boston University School of Medicine in Massachusetts. They have been particularly prominent in drawing attention to the realities, notably the concept of alcohol in moderation being actually good for you.

Ellison was one of the voices heard in 1991 on the television program *60 Minutes*, in which moderate consumption of red wine was touted as the basis of the French Paradox phenomenon: It was the wine that was countering the blocking of arteries (atherosclerosis) by bad cholesterol arising from the fat-filled diet (and the French are rude about English food![5]). The red wine lobby has never felt obliged to be reluctant about these claims and point to a molecule from the grape called resveratrol as being the key component.

Evidence has been mounting that it is far from being this straightforward. The key studies on resveratrol were done in animal systems with relatively huge doses, equivalent to around 100 bottles of red wine every day. You would at the very least need to be exceedingly wealthy and also have the fortitude of a whale to consume that amount. Fortunately, one would not need to, for the simple reason that the growing realization is that the active ingredient is actually plain old ethanol; that is, alcohol. And it does not matter if it is red wine, white wine, beer, or another personal favorite of mine, Southern Comfort. They all, on a same-amount-of-alcohol basis, appear to confer a similar reduced risk of developing coronary heart disease. And it seems to me that a fair guideline to those reasonable quantities would be 26 units per week for a man or 21 units for a woman, with no more than 4 units in any one day, as laid down (so to speak) by the UK government.[6] And that would need to be every day, as observed by Dr. Kenneth Mukamal and his colleagues at the Division of General Medicine and Primary Care of the Beth Israel Deaconess Medical Center in Boston, Massachusetts. Better to have one or two daily than one or two on just a couple of days per week. And no storing it up for the weekend—that is binge drinking.

Nevertheless, people say to me, you have only to look at those who drink wine to see that they are healthier than those consuming beer. To which I reply, "Of course. To buy wine at those ludicrous prices one has to be wealthy, and therefore you can also afford the best health care." Just as tongue-in-cheek I would also say that folks who drink wine eat spinach and go to the gym, whereas those drinking beer sedentarily watch ball games while stuffing sausages and of course those inevitable pretzels. Dr. Morten Gronbaek from the National Institute of Public Health in Denmark studied what people were buying in Danish supermarkets. Those with wine in their baskets were buying more olives, fruit and vegetables, poultry, cooking oil, and low-fat cheese and milk than were beer buyers, while those toting six packs of beer were also hauling more ready-cooked dishes, sugar, cold cuts, chips, pork, butter or margarine, sausages, lamb, and soft drinks. I flippantly suggest that it is only the beer that is keeping these people alive.

I feel that brewers have much to do to convince those who now would automatically gravitate towards wine as their alcoholic beverage of choice to realize that in beer they have a drink of sophistication and genuine consistent quality. Furthermore, there is a vastly greater diversity in beer styles and choices than obtains in the world of grape-derived alcohol (see Appendix B, "Types of Beer").

While they are at it, they might also like to remind the drinker that the evidence clearly weighs in favor of beer in regard to its nutritional merit.

Just the other day, a delightful young nutrition major on campus came to chat with me about class. Answering that she was loving it, she said, "I always smile when I think that if all that was left in this world to drink was water or beer, then it would be far wiser to drink the beer, because it has nutritive value." It would also be free from pathogenic bacteria, which would unlikely be the case for much of the world's water in such a doomsday scenario.

Under those circumstances, of course, the choice would indeed be a no-brainer (indeed, if there really was a global situation where there was nothing else to consume, then the beer sure would be welcome to drown the misery). No right-minded individual, of course, would ever argue in a normal society for overtly considering beer as a nutritive supplement. As we have seen, most of us would for the most part drink it on hedonic grounds. However, it is equally incorrect for the naysayers to suggest that beer is nothing more than empty calories.

Beer contains B vitamins. In one study, my colleagues Andy Clifford and Janel Owens from the Department of Nutrition and I compared the level of folic acid in beers and other beverages.[7] Finding none in wines, we showed that beers differed significantly in their folate levels, with up to 6.1 percent of the recommended daily allowance.

Beer contains substantial levels of polysaccharides, especially the so-called arabinoxylans, which may comprise soluble fiber. Furthermore, we are questioning right now whether some of the molecules that are made in the malting and brewing processes by the breakdown of other complex polysaccharides (called beta-glucans) might be prebiotics, molecules that constitute substrates for beneficial organisms in the large intestine.[8]

We and others have demonstrated the very significant levels of antioxidants in beers. Colleagues at Brewing Research International and Guy's Hospital showed that one of these antioxidants, ferulic acid,[9] entered the body more efficiently from beer than from tomatoes. Alas, this is one of the few studies that have categorically shown the entry of antioxidants from any source into the body, and so there are very real doubts expressed by many about the merit of the very high antioxidant levels reported in products such as wine. It is all very well measuring phenomenal antioxidant *potential* in a product whether it be red wine, green tea, or even beer, but quite another to conclude that it has any merit. It is only if the stuff gets to the parts of the body that matter that there is any possible benefit. The polyphenols in red wine are very complex, relatively big molecules that don't only get snaffled by proteins in the mouth (think bitter and astringent) but also find it a challenge to get through membranes and into the cells of the body. At least in the guise of the much smaller ferulic acid we can contemplate some possibilities, and that is a molecule in substantially higher levels in beer than wine.

These arguments of course apply to all the components of any foodstuff, including beer. We must always make claims with caution, such as in the case of our relatively recent work on silica (see Chapter 6, "Despite the Odds: Anti-Alcohol Forces"). Casey and I found that beers could contain up to 56 mg/L silicate. There is no recommended silica intake, but the average consumption in the United States is 20-50 mg per day. So at the highest level we measured, a 12-ounce serving of the beer would contribute all of the daily intake at the lower consumption rate. Assuming, that is, it gets into the body. But it seems that the silica in beer is in its most accessible form, and indeed the esteemed British medical scientist, Dr. Jonathan Powell, has very clearly drawn a correlation from his research between beer consumption and reduced risk of osteoporosis. The caveats discussed earlier in this chapter naturally apply.

Osteoporosis is of course but one affliction associated with advancing years. And it is not the only one that has been claimed to be countered by moderate consumption of beer. It has been suggested that the risk of Alzheimer's disease is lowered. It is indicated that cognitive function is improved.

Quite whether there is a direct causal link is uncertain. For example, there are those who believe that aluminum is associated with Alzheimer's and that the sequestration of aluminum by silicate—such as that found in high levels in beer—helps counter the risk. Equally possible (or additional) causes are to do with relaxation. The mellowing impact of a daily drink, associated perhaps with convivial company, may have a major beneficial role. As Professor Robert Kastenbaum says

> There is by now sufficient information available to indicate that moderate use of alcoholic beverages is pleasurable and beneficial for older adults. The lower-ethanol beverages such as beer and wine may have their place in the lifestyles of some older adults.

Beer has been linked to the reduced instance of kidney stones and gallstones. Beer is rather more diuretic than is water, and fresh beer makes you pee more than does stale beer!

We have already peeked at atherosclerosis, the alcohol favorably affecting the balance of "good" versus "bad" cholesterol in the body, and also reducing the risk of blood clotting. Drinking has been linked to increased blood pressure, but the current belief is that the blood pressure of nondrinkers is in fact *higher* than in those consuming 10-20 g alcohol per day.[10] Hypertension certainly increases the risk of stroke, but again it seems that there may be a lowered risk of stroke for moderate drinkers and only when drinking is heavy (more than six drinks per day) does the risk of stroke become significant.

There is now good evidence that stomach and duodenal ulcers are induced by a bacterium called *Helicobacter pylori.* This organism is inhibited by alcohol—so there are reports of reduced chance of ulcers through moderate drinking.

All is not rosy for drinkers. The risk of pancreatitis is increased in heavy drinkers. Some beers should be avoided by sufferers of gout because they can contain significant quantities of purines. Hangovers are likely caused by a buildup of acetaldehyde, a breakdown product of alcohol.

On the other hand, moderate drinking has been linked to reduced risk of developing diabetes.

Perhaps the most emotive discussions revolve around cancer. A quarter of all deaths in developed countries can be ascribed to cancer. Cancer occurs when cells divide uncontrollably, perhaps invading other tissues. The DNA becomes damaged, leading to mutations that cause those cells to escape from the control processes that keep them in check. Anything capable of mutating DNA can in theory at least lead to cancer.

Some claim that alcohol is a mutagen, but this is certainly not the case. But are the metabolic by-products of alcohol mutagenic? Or are there traces of other mutagenic components? There is no firm evidence one way or the other.

It is now generally accepted that obesity increases the risk of developing cancer. Is the link truly causal, or do people who overeat consume foods that contain carcinogenic substances? Taking the relationship at face value, anything that imparts calories will contribute to the risks if consumed to excess, including alcoholic beverages.

In many societies the beer drinker is also the smoker or, even if they are a nonsmoker, someone who frequents pubs where others are billowing out fumes. Uncritical evaluation might reveal a correlation between beer drinking and cancer, when in reality the causative agent is burning tobacco.

The National Institute on Alcohol Abuse and Alcoholism (NIAAA) draws attention to the tremendous inconsistency in data. They caution against *abuse* of alcohol, and also draw attention to the claims that alcohol potentiates the damaging impact of the prime carcinogen, notably tobacco.

Arimoto-Kobayashi from Okayama University discovered pseudouridine in beer, a substance capable of protecting against mutagens. Pseudouridine seems to account for only 3 percent of the antimutagenic complement of beer, which leaves quite a lot still to be discovered in this area. Perhaps some hop-derived compounds are important, of the type suggested by scientists at Oregon State University. It is always wise to be cautious in claims and aspirations, especially with such an emotive and potentially devastating issue such as cancer. It would be dreadful to lead those suffering from the ailment to false hope, just as it would be irresponsible to overplay the risks of alcoholic beverages for this disease.[11] It really does appear that the drinking of beer within recommended limits (such as those recommended by the United States Department of Agriculture) will not manifestly increase the risk of contracting cancer. Despite the optimism that there may be ingredients of beer that may have anticarcinogenic value, we really must see where the accumulation of evidence takes us.

❋ ❋ ❋

So I wrote to the author of the diet that had become all the rage a few years ago. I was incensed. He had written that beer was the worst thing you could put into your mouth, because it contains (or rather he *said* it contains) maltose, and maltose was the worst of all sugars for the glycemic index.[12] I asked him (politely) if he had ever heard of the interesting phenomenon known as fermentation. I pointed out that in this process maltose is converted into alcohol. Sure, alcohol is the major source of calories in any alcoholic beverage.[13] But it isn't maltose. Now the diet says "due to recent research we can rescind the ban on beer."

Bizarre. Nonetheless there remains the myth about the beer belly. The truth really is rather straightforward: It's a case of calories-in versus calories-out.[14] There is nothing peculiar about the calories in beer: If they are counted among the daily calorie intake, provided the latter level is in balance with (or less than) the calories burned off, then there will be no body fat accumulation, in the belly or anywhere else for that matter. And if you compare a Bud at 150 calories with a slice of lemon meringue pie with at least *twice* that amount then it is a no-brainer!

❋ ❋ ❋

Certain components of beer may induce symptoms in sensitive individuals—for example, proteins claimed to be deleterious for sufferers of celiac disease. Medical advice for such patients is to avoid foodstuffs derived from wheat and barley. Therefore there is interest in beers such as Redbridge that are brewed from sorghum and that do not contain sensitive proteins.

Some men have heard scary stories about hops being rich in phytoestrogens, but they should relax: The amounts getting into beer are minuscule.

Finally, let me make mention of the old adage that nursing moms should drink beer for good breast-feeding. This certainly has nothing to do with female hormones. Some believe that there is a carbohydrate in beer coming from barley that promotes secretion of the hormone prolactin, which encourages milk production. It may be that the reality is that the beer is nothing more than a reward after a somewhat demanding experience. Just another example of beer's role in a holistic lifestyle.

9

Whither Brewing?

The worm turned in 2007. A fire in a hop store in Yakima in 2006 hadn't helped, taking with it substantial reserves. But the real cause of the hop crisis was founded on many brewers' reluctance to pay the hop folks quite the price that is perhaps warranted for this unique crop. As one brewer said to me: "By the time we have finished with them, they don't have a pot to piss in." Some hop growers had had enough and grubbed out their yards, to be replaced with more reliable and rewarding sowings. Meanwhile, although the average bitterness content of the world's beers was trending downwards (despite the best efforts of some hop-heads in the US craft sector), the world-wide production of beer was ever-growing. The tipping point had been reached, and there were not enough hops. Far-sighted brewers who had contracted ahead and who also had a reputation for treating hop merchants honestly and fairly were okay. But many others were hurting and desperate for supplies. The price zoomed.

Imagine what will happen if and when the hop people succeed in finding alternative outlets for their product. And they are looking. Already they are selling resins into the food industry for their preservative properties. Hops contain a diversity of fascinating molecules. According to Denis De Keukeleire of the University of Ghent:

> *The diversity of natural hop constituents undoubtedly accounts for the varied and rich panoply of bioactivities hitherto reported: sedative, anti-stress, soporific activities, estrogenicity, treatment of complaints related to the menopause, anti-cancer properties (in particular inhibition of hormone-dependent cancers of breast, uterus and prostate), bacteriostatic activity, anti-inflammatory action, stimulation of the digestive tract, diureticum and agent against bladder complaints, diaphoretic and perspiration-stimulating effects, and anaphrodiasiacum.*

The most potent phytoestrogen yet identified has been found in the hop, so small wonder that another outlet is for breast enhancement products. Others genuinely see potential for hop-derived candidates in anticancer modes. The list will grow.

And as more and more high-value outlets develop for hops, so will it demand either more hop acreage or higher prices, or both. It is a poignant example of the sensitivity of the brewing industry to pressures on raw materials.

The stresses are no less acute for malting barley. Within the current brewing paradigm it is not a case of any old barley. Brewers demand the best, in particular those grains that yield the highest levels of starch. More starch means more sugar generated in the brewhouse, which in turn means a higher yield of alcohol after fermentation, or, in other words, more beer. It is actually more complex than that, for the barley must also yield its starch readily, not cause problems in the brewhouse, must not be associated with adverse flavors, must be resistant to infection and infestation...the list goes on. To ensure high starch, the barley must be limited in protein, which in turn means that nitrogenous fertilizer application to the growing crop must be limited. That means less yield. And so, to persuade the farmer to grow such barley, there needs to be an extra reward, in the shape of a so-called "malting premium." But just in case the crop falls short of the stringent demands of the maltster and thence the brewer, then the premium will not be paid, the crop will be rejected, and all the farmer can hope to do is sell it off for feed, at a vastly reduced price. Small wonder that many will choose not to run the risk and elect to raise a more reliable crop, such as corn or wheat. Indeed I recently heard Rob McCaig from the Canadian Malting Barley Technical Center say that malting barley is a measly eighth in net return per acre in Canada after lentils, canola, yellow peat, oats, flax, amber durum, and wheat.

Barley (just as the hop) is a living organism prone to seasonal and geographical variation. In the world of wine, year-to-year variations are tolerated, the product changes, and the wine is labeled according to its vintage. In brewing there is not such acceptance. Brewers demand consistency and predictability. To an extent they are prepared to adjust their processes to overcome inherent and unavoidable variation...but only up to a point. The more consistent the raw material, the more straightforward the brewer's job becomes.

The balance between malting barley availability and demand by brewers is, should we say, "tight." In years where parameters such as protein are trending high there is inevitably a threat of shortage for some (again, forward contracting works best, but not all brewers have the necessary

infrastructure, strength, or aptitude). Moreover, there is a
geographical issue: The growing beer markets do not coincide
with the premier barley growth locations, such as Canada, the
US, Australia, and Europe.

Consider now the third of the four great brewing ingredients, water. (We need not worry about yeast.[1]) Brewing
demands a lot of water. Not only is it needed in the beer itself
(most beers are in excess of 90 percent water), but it is used in
substantial quantities for cleaning (especially in the packaging
hall) and to raise steam in the boiler house. A very few years
ago, the best of brewers in the world were using five or six
times more water than found its way into the packaged beer.
Now they are working toward a ratio of 3.5:1 or less. As I
write, the world leader is the Yatala brewery of the Foster's
company in Queensland, Australia, with an astonishing ratio
close to 2 to 1.

Fundamentally there are three strategies by which a
brewer may do its thing for the water effort: Use less, recover
as much as possible, and reuse after treatment. The main gen-
erator of waste water is the packaging operation, especially
when beer is filled into cans and bottles. Bottles can be of two
types: so-called "one-trip," which means they are nonreturnable, and returnable. The customer sends the nonreturnable
bottles to recycling—and I have done a good job for my local
Episcopalian church by donating my empties so that they can
get the 5 cent redemption value. Drinking for God, you might
say. Such non-returnable glass, in which the glass is eventually
melted down and reformed into new bottles, is much more
water-friendly than those returnable bottles that get sent back
to the brewery. The brewery packaging hall is obliged to sort
the latter (can't have greens and browns together, for instance)
and then subjects them to a mighty cleanup involving copious
water and cleaning agents. Think of that the next time that

you are sitting with such a bottle and doing the inevitable thing with your chip bag of folding it up tightly and pushing it into the bottle. Somebody is going to have to extract it again.

In North America, it's pretty much a case of returnable glass in Canada and one-trip in the US.

It is not only a matter of water used for cleaning bottles (and empty cans as they arrive in the packaging hall). A pasteurizer uses much water. Conveyors demand lubrication. The floors need to be scrubbed. And much more besides.

Reduction of water consumption at earlier stages in the brewery is a trickier prospect. The brewer might shorten the boiling operation, while being ever-mindful of the impact that this can have on product quality. They might clean less frequently, again being mindful of jeopardizing standards. And moves that some brewers are making to have continuous processing would certainly help here.

Conserving water, though, starts with good housekeeping, and so brewers should be ever-mindful of leaking valves, unnecessarily running faucets, and the like.

Ultimately, saving water as well as most other improvements that can be made in a brewery comes down to gaining the commitment of the number one resource, people.[2]

When we speak of matters ecological, it is fashionable to refer to the carbon footprint. A carbon footprint is a measure of the impact our activities have on the environment and thus climate change. It relates to the amount of greenhouse gases (notably carbon dioxide and methane) produced in our day-to-day lives through burning fossil fuels for electricity, heating, and transportation, and so on.

When it comes to the brewing of beer, we must raise the question of where do we start and end?

A true carbon footprint surely embraces all activities from raw material production and transportation through to distribution and even consumer practices.

Several studies have been undertaken on the carbon footprint of brewing. The New Belgium Brewing Company declared what the footprint is for a six pack of Fat Tire Ale.[3] Other studies have delved even deeper, factoring in for instance the contribution made by employees traveling to work.

Naturally, there is a strong correlation between energy consumption and carbon footprint, carbon dioxide being generated when fuel is burned. So any measures to conserve energy will have a direct impact on the carbon footprint bottom line: turning off lights; energy conservation protocols in all operations; not forgetting to get employees to think about the most efficient way they can get to work.[4]

And then there is waste. Most brewers divert their spent grains to cattle feed—quickly, otherwise they sour, much to the distaste of the cows. Surplus yeast can be used to feed pigs—the porkers being especially delighted as there is usually plenty of alcohol still bathing the cells. Spent water is increasingly diverted by brewers to anaerobic digesters, where bacteria convert the molecules in the water to methane, which can be burned with the release of utilizable energy. Some brewers put waste water onto fields, to grow crops such as canola that can be converted into biodiesel to fuel trucks. Some are debating channeling gases from landfill to burners. More than one has installed vast solar panel arrays; one I know is aiming to erect a wind turbine.

Radical Solutions

Most brewers are pretty conservative. I think it was a Busch who coined the phrase "if it ain't broke don't fix it." To suggest, therefore, that they pursue fundamentally different ways of doing business is hardly to court popularity. Yet there are moves afoot to think radically.

Right now, the folks that sell enzymes to brewers are urging their customers to do the decent thing: Think of the planet, think of the abundant water and energy that making malt demands, as well as all the greenhouse gases spewed out, so don't make the enzymes by germinating barley. Rather use raw barley, together with the enzymes from bacteria and fungi that they (the enzyme company) will readily supply you with.

There is nothing technically novel here, although the marketing push is clever indeed. The problem remains one of taste. Malt makes a meaningful contribution to the flavor of many beers. Barley is, well, grainy and astringent. One would have great difficulty trying to match the taste and aroma of a given brand made on the one hand from malt and the other from raw barley.

However, for *new* products, the concept is certainly legitimate. Think happoshu and Third Way drinks in Japan, where the driving force is savings in tax.[5] Make a product cheap enough, and there will be consumers willing to buy it, especially if there is a "green handle" as well.

In my lab we actually made a "beer" by adding flavor, color, and foam to a batch of vodka, and the resultant drink was clearly accepted as a beer (if not a particularly good one) when presented to a panel of tasters. When we told them how it had been manufactured, it did not affect at all their liking or disliking scores.[6]

The legendary Fritz Maytag says, "A beer has to taste right here" (gesturing to his heart) "as much as it does here" (pointing to his head). And, frankly, the ersatz beer to me felt better in the mouth than it did in my heart. But as our understanding of what determines the intricate complexity of beer flavor continues to improve, surely products made in this way will only get better and better.

But is it beer? Does it matter? What does the customer want? Certainly it would make eminent sense to make a malternative in this way were it not for the fact that it would be taxed at a much higher rate.

For me, a beer is more than merely a long, cold, and slightly alcoholic drink. It is more spiritual than spirit, as we will now explore.

10

God in a Glass

Ralph Waldo Emerson (1803-1882) was a Bostonian Unitarian minister. His scholarship led him away from his church towards transcendentalism. He was one of several people seeking to come to terms with traditional religious teachings in the context of the new age of which they were a part, an age of burgeoning scientific understanding and rationalizations based on logic as opposed to belief. Emerson, a graduate of Harvard, began to read Hindu and Buddhist scriptures, and started to compare his new awareness with what he had assimilated in his Christian upbringing. He realized that spiritual truth could take many forms—millions of people following an alternative tradition must surely mean that there is another form of the same "truth." Emerson said:

> We will walk on our own feet; we will work with our own hands; we will speak our own minds.... A nation of men will for the first time exist, because each believes himself inspired by the Divine Soul which also inspires all men.

The transcendentalists confronted issues of slavery and the rights of women, believing that all of humankind could be simplified to the level of the soul and that all people should be able to reach out to the divine.

Emerson also penned the immortal words, "God made yeast, as well as dough, and loves fermentation just as dearly as he loves vegetation."

In a sentence, the transcendentalist linked a higher power with the daily reality of fermentation. And as a good Bostonian he will have been well aware of beer, and that yeast was the catalyst to its production, even though it hadn't been until the work of the likes of Pasteur, Cagniard-Latour, and Schwann,[1] that the true role of yeast acting as a living agent was demonstrated. It was long after Emerson's death that the first enzyme was extracted from yeast by Buchner in 1897.[2]

The conversion of grain to the feedstock (wort) for yeast to turn into an alcohol-containing triumph flavored with hops to delight humankind is surely a sublime example of a profound Buddhist concept. In *The Tibetan Book of Living and Dying*, Sogyal Rinpoche tells of a Buddhist scripture that speaks of "conditionality." In addressing the matter of rebirth, the author describes the concept of *successive existences*:

> The successive existences in a series of rebirths are not like the pearls in a pearl necklace, held together by a string, the "soul" which passes through all the pearls; rather they are like dice piled one on top of the other. Each die is separate, but it supports the one above it, with which it is functionally connected. Between the dice there is no identity, but conditionality.

Sogyal Rinpoche invokes the tale of a Buddhist sage, Nagasena, explaining the concept to King Milinda. One of the examples Nagasena uses to illustrate the idea is the conversion of milk to curds and ghee: They are not the same as milk, but nonetheless are dependent upon it.

The example could just as easily be beer. Glasses of Budweiser or Sierra Nevada Pale Ale do not resemble one another in appearance or taste, and even less so do they resemble the raw materials from which they are made. But the living essences of grain and hops and yeast, bathed in purest water, are there. The beer is conditional upon the preexistence of its raw materials and on the journey that they took in field, malt house, and brewery.

Think of this: acre upon acre of soil in Montana, newly plowed and sown with barley seed from an earlier crop. Caringly, the farmer irrigates the fields as a golden sun beats down. The tiny shoots break through the soil level and wind their way towards the heavens, basking in the sunlight, inhaling the carbon dioxide and stocking the larder of the grain with carbohydrates. Below ground the tiny roots get stronger and stronger as they burrow through the earth, gratefully drinking nature's goodness from the terrain. Months later the barley is fully grown, and the grain now waves proudly at the head of each stalk, bravely enduring the mighty harvesters that separate it from the mother plant and usher it to the vast dark silos that will be its next home.

Soon vast numbers of kernels will be plunged into cool clean water, thirstily satisfying the need of the baby plant and moistening the starchy food reserve that the embryo depends upon to support its germination. Slowly and surely, the myriad complexity within each barley kernel makes for the partial digestion of the food reserve, and the nutrients enable the fledgling rootlets and shoot to develop. In a week or so, the grain is helpfully softened and has generated within it the enzymes that will now be able to break down the starch in the brewhouse. So the maltster halts the germination: He toasts the grain and, their work done, the embryos give up their brief lives. The heating causes new changes: Colors get darker, different flavors develop. Sometimes the cooking is gentle—giving pale lemon hues, then ambers and gentle sulfury flavors. Such malts will be transformed to lagers. Or perhaps the kilning is more intense, leading to reds and lighter browns, and toffee- or nutlike notes revered in pale or darker ales. Other grains are roasted, yielding dark browns and blacks, with burnt tastes of mocha, chocolate, and coffee. Such malts have sacrificed themselves to stouts.

Meanwhile in a garden somewhere in Oregon, Idaho, or Washington, hop vines wind their way over trellis work stretching as far as the eye can see. Come the fall, beauteous cones adorn the female plants and deep within each lie rich yellow lupulin glands suffused with the resins that will make a beer bitter to the taste and oils that will afford the triumphant "nose" to ales and lagers.

Taste the barley or the hop as they have matured in the field, and you will not sense delight. But transformed, respectively, by malting and by the boiling process in the brewhouse, then we see the magnificence of our God's deliverance.

The starch broken by enzymes in the brewhouse is thereby rendered in the form of sugars that satisfy yeast. And so, after the boil, the liquid that is wort is cooled and yeast added. Greedily the yeast consumes the sugars, turning them into alcohol and carbon dioxide, and extracting sufficient goodness to allow the simple cells to grow and produce daughter cells. The yeast multiplies, and after several days there is perhaps three times more yeast than at the start—and beer whose flavor is so much more pleasing than the "wort" that the yeast had been fed. And what wondrous economy: Some of the new yeast can be used to "pitch" the next fermentation (no wonder the medieval brewers called this stuff that they saw collecting on their fermentations "godesgoode"). And the rest of the yeast can go to feed animals, even humans as Marmite or Vegemite,[3] just as the parts of the malt that weren't extracted into wort head off to delight cattle.

Nothing wasted. Barley (plus other cereals like wheat, corn, and rice), hops, yeast, and water transformed by miracles of metabolism, fed by nature's bounty.

Conclusion

Over the years that I have lived in California, I have come to terms with the neo-prohibitionist tendencies displayed by rather a large number of people. Time was when it would irritate me: "Shit, you get on with your life, and I will get on with mine." Now I smile and accept: We are each of us driven to declaring with various degrees of intensity our beliefs and understandings of what is true, what is important, what is just and right. I have come to realize, though, that we should each respect another's truth (even dogma) but to fully expect that they should show consideration for ours. Beer is not for everyone, but please do not expect that it is something that I should forego for the simple reason that it does not sit comfortably with you.

The arguments against alcohol either center on a religious teaching or alternatively a thesis that alcohol in any amount or circumstance is antisocietal. The former is a stance to be respected and understood as the dogma of a given religion within which, hopefully, you have the free will to participate or not. The latter, however, eschews the very real benefits and positives that literally flow, for example, from a beer tap and seeks to stamp out the freedom that an individual should have to indulge in an element of this life that for the longest time has brought contentment, community, and (it is increasingly recognized) healthful comfort to those who indulge respectfully. As such, it would be as wrong to seek to ban beer, or make it prohibitively expensive, or marginalize it as it would be to forbid skiing because limbs can be broken, or driving because some people motor far too quickly, or candy because it leads to obesity and thereby all manner of physical crises. Equally it would be no less and no more justified than seeking to eradicate or diminish the religious belief of another.

I recall as an 11-year-old my very first game of rugby,[1] for here was my first exposure to religious bigotry that made Wigan and its environs a low-key version of Northern Ireland, which is not that far away across the Irish Sea. That opener was down the road at the St. John Rigby School. The Catholic team kicked off and the ball looped into my arms, at which time I heard a priest on the touchline bellow "get the little Protestant bastard." Sure enough I was hammered, emerging stunned from beneath a pile of bodies to find that my shorts were flapping about me as they had been rent asunder.

Although meaningful for me at the time as I dragged myself back onto trembling legs, it is a mere trifling example of the evils that surface again and again in this imperfect reality that we call our world.[2] Not long ago the Israeli army invaded the Gaza Strip in response to rocket attacks on their country from a regime that does not believe in the right of the Jewish state to exist. It is less than a decade since the nightmare of 9/11, founded on religious extremism; a fanaticism spawned a millennium ago with the crusading Christians seeking to eradicate their Muslim foes. In India today there are countless conflicts between Hindus, Muslims, and Christians.

Naïveté, maybe, but I will never fathom how religion can be a true basis for conflict, for surely all religions hold at their heart a message of love and peace. And yet there is such dispute between faiths and even within doctrines—witness the recent chasms in the Episcopal Church founded on an intolerance of homosexual and female clergy.

Is it so very different for some to be seeking to deny me the quiet pleasure of my pint? For surely as long as I am peacefully enjoying the beer, nonexcessively and without harming myself or others, it is a simple and innocent aspect of God's reality? Or as C.S. Lewis wrote

> *An individual Christian may see fit to give up all sorts of things for special reasons—marriage, or meat, or beer, or cinema; but the moment he starts saying the things are bad in themselves, or looking down his nose at other people who do use them, he has taken the wrong turning.*

So in my beery world may I tolerate those folks who like their beers smothered in hoppiness just as I would hope they would tolerate the skill devoted by the big brewers to making bland lagers so consistently well. May I tolerate those who rejoice in beer exposed to bright sunlight so that it reeks of skunks while begging them to acknowledge my taste for tepid flat ales in an East End pub in London. And may the wine drinkers of this world know that I am totally tolerant of their preferred beverage just as I want them to allow me to sing the praises of my own.

And likewise may each of us tolerate black and white; straight and gay; rich and poor. Let us recognize that the selfsame humanity resides in a president and a panhandler, in a CEO and a janitor; in man, in woman, in child.

Endnotes

Preface

1. *Beer: Tap Into the Art and Science of Brewing.*
2. And my friend Michael Lewis wonders *why* I write so much!
3. On February 12, 1972, I was told to pitch up at Cleminson Hall, Cottingham (the largest village in England), because a girl was eager to meet me, somebody who had seen me at a Hull University hall-of-residence dance and who was entirely smitten. (It was only weeks later that she and I found out that this was an entire fabrication and that the same story in reverse had been sold to her.) To prepare myself for the ordeal I took in the pub—and several pints—and, fortified also with a number of cigarettes, I allowed myself to be introduced to Diane Heather Dunkley. I vaguely remember what she was wearing—a sensible blouse and a tartan skirt. With more certainty I can see my attire—vivid orange vest, purple trousers with stitched-on black flares, huge

zip-up boots, topped off by hair to my shoulders, side-burns that I could tie under my chin, and black horn-rim glasses. Factor in the beer and cigs, and you will recognize that I was some catch. But I was lukewarm. Heck, scattered on her desk were copies of *The Ringing World*—here was a chick who pulled more bell ropes in cathedrals and churches than she did men. Our conversation was amicable enough, though unspectacular. However, things went well enough for her to agree to join me in the pub next evening. She saw a very different Charlie—I had toned down my clothes and was stony cold sober when we met up. In turn, she had livened up a bit and was wearing a smock and jeans. I remember thinking that she was Ali McGraw meets Karen Carpenter. We have been together ever since. And she truly is young at heart and in appearances. Only recently she was asked, "and what is your relationship to the professor? Is he your father?"

Introduction

1. **Bitter**. Pale ale served on draft dispense.
2. **Blue bag**. In those days, the concept of "ready salted" was unknown, the technology not being available to ensure an even coating of salt (sodium chloride) on every crisp (chip). Upon opening a bag, one had to search for the diminutive blue packet, which comprised a small amount of salt wrapped in a tiny piece of blue paper. We felt ourselves winners if a packet contained two or more salt parcels.
3. **Scratchings**. Pork rinds deep-fried and served dry at room temperature as a snack.
4. **Cockles, whelks, and mussels**. Types of shellfish. Cockles are similar to clams, whelks are sea snails.

5. **Sally Army**. Salvation Army, the newspaper for which is *The War Cry*.

6. **Arthur Koestler**. Naturalized British essayist (1905-1983) of Hungarian birth.

7. **St. Thomas' Church**. Anglican church in Up Holland, dating back to a pre-Reformation priory first established in 1307 (www.stthomasthemartyr.org.uk/history. htm). Up Holland is a village about three miles west of Wigan in what (at the time of the author's boyhood) was the Red Rose county of Lancashire. Perhaps Up Holland is best known for its Roman Catholic seminary.

8. **Rugby League**. At that time the professional variant of Rugby Union, one with 13 players per side as opposed to 15.

9. **Mild**. Brown ale on tap.

10. **Jubilee, Mackeson**. Sweet stouts, sometimes called Milk Stouts.

11. **Bass No. 1, Gold Label**. Barley wines.

12. **Wigan Pier**. A coal-loading staithe by the Leeds-Liverpool Canal, a 127-mile long canal linking the port of Liverpool with the industrial heartland of Yorkshire. It was immortalized by the music hall comedian George Formby, Wigan-born and bred, who would make a play on the use of the word *pier* even though industrial Wigan was hardly a resort. Formby's son, also George, became even more famous as a morale-boosting, buck-toothed ukulele-playing comedian and film star in the Second World War era.

13. **Ribble bus**. Red-liveried bus company based in Preston but which merged into Stagecoach in 1988.

14. **Smoke-free zones**. A smoking ban in England making it illegal to smoke in all enclosed public and work places came into force on July 1, 2007.

15. **Dark Satanic Mills**. From a short poem by William Blake (1757-1827) and immortalized within the hymn *Jerusalem*, music by Sir Hubert Parry (1848-1918):

And did those feet in ancient time.
Walk upon England's mountains green:
And was the holy Lamb of God,
On England's pleasant pastures seen!

And did the Countenance Divine,
Shine forth upon our clouded hills?
And was Jerusalem builded here,
Among these dark Satanic Mills?

Bring me my Bow of burning gold;
Bring me my Arrows of desire:
Bring me my Spear: O clouds unfold!
Bring me my Chariot of fire!

I will not cease from Mental Fight,
Nor shall my Sword sleep in my hand:
Till we have built Jerusalem,
In England's green & pleasant Land.

There are those who would have this as the National Anthem—witness the Last Night of the Proms.

Chapter 1

1. **Endowment**. Anheuser-Busch donated a significant six-figure sum to the campus in the late 1990s. This was invested and the interest generated annually is in part reinvested to grow the fund, with the balance coming to me as the endowed professor to spend wisely in support

of the brewing program. The activities are also supported by the generous donations of many other brewing and supplier companies across the world.

2. **Aggie.** Alumni from UC Davis are Aggies, speaking to the campus's agricultural roots.

3. **Tasting.** I vowed that day that I would never again pass judgment in that room. Every time thereafter that I was invited to taste I sat firmly on my hands and declared everything excellent (as indeed it always was). Come the day that Doug had a glass of one beer placed in front of me but nobody else. "As a special favor to me, I would ask you to please comment on this one." I protested, but he prevailed and so I smelled the product and tasted it. It was awful: grainy, worty, astringent. I gulped and gave my honest assessment. He smiled. "Good, because that's...." and he named the beer as being that of a major international competitor.

4. **Scottsdale.** The event coincided with me working on a book by Michael Brown called *The Presence Process*, basically an exercise in living in the now and making decisions in the here-and-now rather than on "issues" spawned in the past or those that might (or might not) occur in the future. The journey Brown advocated for inner awareness involved abstaining from alcohol for ten weeks. I did, successfully and relatively painlessly— but quite a challenge when one joined the big wigs from Anheuser-Busch for steaks in Arizona. "Five Buds and a glass of water, please."

5. **Wine versus beer.** *Grape versus Grain*, Cambridge University Press.

6. **A-B investments.** These included major shareholdings in Modelo (Mexico) and its Corona beer; in Tsingtao, China; and the Stag Brewery in Mortlake, London, England.

7. **Anheuser-Busch InBev** is the world's third largest consumer products company with operations in more

than 30 countries. It has some 300 brands, brewing four of the top-ten selling beers in the world, and employs approximately 116,000 people.

8. **Beer volumes**. A barrel of beer (US) comprises 31 gallons. The international unit for volume is the liter or the hectoliter (100 liters). One gallon equates to 3.7853 liters in the US; therefore a US barrel holds 1.1734 hectoliters (hl).

9. **Chopp**. Means "draft."

10. **Skol**. "Cheers" in Scandinavian countries, derived from the original term *skál* or *skål* for a drinking cup.

11. **Common brewer**. The first brewing companies established to produce beer at increasingly sizeable breweries separate from retail outlets.

12. **Red Barrel**. A much-maligned beer brewed by Watney's. One of the first major kegged beer brands; that is, brewery-filtered and pasteurized products that moved away from the traditional cask ales of England and Wales. Designed for stability, convenience, and for the Briton traveling abroad.

13. **Scottish & Newcastle**. 1749 saw the founding of William Younger's Brewery in Edinburgh, Scotland. A merger with McEwan's in 1931 led to Scottish Brewers, and then in 1960 there was a new merger, with Newcastle Breweries, to yield Scottish & Newcastle. In 1995, with the Thatcher Beer Laws edict in full swing and other big breweries shifting out of production, they acquired Courage to become the UK's leading brewer, having previously been only number five. They went on international missions, acquiring France's Kronenbourg in 2000 and Finland's Hartwall in 2002, a company with major interests in Russia, Ukraine, Kazakhstan, Latvia, Lithuania, and Estonia. In July 2003, they bought the cider company Bulmer's. The company started rationalizing its assets, closing the historic Fountain brewery in Edinburgh and the Tyne Brewery in Newcastle in 2004.

And then in 2008 Heineken and Carlsberg moved in. Carlsberg took the Finland/Eastern Europe share, while Heineken gained the UK and other European elements. Soon afterwards they announced the closure of the old Courage brewery in Reading, the biggest brewery in the UK.

14. **Raj.** The period of British colonial rule in South Asia (1858 to 1947). It is from this time that the British love affair with the curry began. The author is one such aficionado.

15. **Bass Red Triangle.** On New Year's Eve in 1875, a Bass employee waited outside the registrar's office overnight so as to be first in line to register a trademark. Actually, he filed two, the second one being a red diamond for the company's strong ale. These days both are owned by Anheuser-Busch InBev. The red triangle is seen in many pieces of artwork, most famously in Manet's *Bar at the Folies-Bergère* and numerous pieces by Picasso. The beer was also portrayed in the original illustrations for Kenneth Grahame's *Wind in the Willows*.

16. **Lord Gretton.** John Frederic Gretton, Member of Parliament for Burton-on-Trent.

17. **Percy James Grigg.** Secretary of State for War.

18. **Carling.** Thomas Carling was a Yorkshireman who emigrated to Ontario in 1818. His home-brewed beer was a great local success, and the Carling company was established in London, Ontario, in 1840. Many years later the brewery was acquired by E. P. Taylor's Canadian Breweries Ltd., which duly became Carling O'Keeffe, thereafter merging into Molson, which of course would become Molson-Coors. Thus did the wheel come full circle, for when Bass sold its breweries post-Thatcher's Beer orders, the Carling brand was acquired by Coors.

19. **Cask ale.** Also known as cask-conditioned beer, it is beer racked from fermenter without filtration or pasteurization into barrels, with the addition of priming

sugar, isinglass finings, and quantities of whole hops. The residual yeast uses the sugars to naturally carbonate the product, while the hops provide the classic robust dry hop aroma. The isinglass finings clarify the product. The beer style demands care and attention; otherwise, it can rapidly spoil. The author has been known to claim it to be the most drinkable beer on the planet and likens it to an angel weeping on one's tongue.

20. **Marston's**. Brewers of the famous Pedigree ale, long-prized for being fermented in a Burton Union system. The company is very different now to what it was when founded, and again the changes speak to the tremendous consolidation of the British brewing industry. Marston's was acquired by Wolverhampton and Dudley Breweries in 1999, a company that also acquired Mansfield Brewery. Thereafter followed Jennings and Ringwood. The company changed its name from Wolverhampton and Dudley to Marston's in 2007.

21. **Burton Union system**. A recirculating fermentation approach from the early nineteenth century comprising a row of casks connected to a "top" trough via a series of "swan-neck" tubes. As the ale ferments, the top fermenting yeast rises into the top trough from which it can be removed, with the beer flowing back into the casks. A version of the system is employed in California by Firestone Walker.

22. **Cylindro-conical vessels**. Designed by Nathan, cylindrical vessels with a conical bottom (CCVs). The vessels have cooling jackets through which cooling liquid counters the metabolic energy of the yeast, thereby allowing the temperature to be controlled with the achievement of a consistent fermentation. Once fermentation is complete, the contents of the vessels can be cooled right down, leading to settling of the yeast in the cone, from which it can be removed.

23. See Appendix A for an overview of the technicalities of malting and brewing.

24. **Franchise brewing**. A very accurate adage is that because of its inherent instability, beer is usually best when consumed closest to the brewery. Not only that, but most beers are at least 90 percent water. It really does not make sense to ship large volumes of beer large distances. Thus a company will either have its own breweries strategically located to satisfy a distribution network, or will seek to have its beers brewed under contract by other companies.

25. **BRF International**. Founded as the Brewing Industry Research Foundation (BIRF) soon after World War II, it was funded on the basis of a research tax on barrels of beer brewed in the UK, the membership fees from the companies being used to support research in a large converted country house at Nutfield in Surrey. In due course the "I" was dropped, making it BRF. In 1991 the organization morphed to become BRF International, as members were recruited globally. Later in the nineties the "F" was lost, making it BRi. Then in 2009, the organization merged with the Campden & Chorleywood Food Research Association to form Campden-BRi. The author had two "tours of duty" at BRF (as he prefers to call it): 1978-1983 and 1991-1998.

26. **Wort clarity**. The boiled extract of malt (wort) may be relatively free from particles ("bright") or may contain many particles ("dirty"). Those preferring the former believe the particles are a negative for product quality. Those preferring the latter believe that the particles promote fermentation.

27. **Wort separation**. Separating the wort from the spent grains is most often achieved using either a lauter tun or a mash filter (see Appendix A). Mash filters give worts with fewer particles which are therefore said to be "brighter."

28. **Fermenter geometry**. Although CCVs (see endnote 22) are tall and relatively narrow, making for efficient packing into a relatively limited space, some brewers

prefer so-called "horizontal fermenters," which might be likened to a CCV on its side (without the conical base).

29. **Yeast contribution**. Although the yeast is important in all beers as the source of alcohol production, it is debatable whether it has a major role in determining the character of those beers that are made with strongly flavored malts and to high levels of hoppy aroma. The author recalls a famous brewery owner once claiming (facetiously?) that the flavor of his beer was all down to the malt and hops and that he had even once used baker's yeast for fermentation.

30. **Water**. It is often claimed that this is the reason why it is impossible to brew a given beer anywhere other than the original brewery. This is patently untrue, for the water can be adjusted to have an identical analytical value.

31. **Rocky Mountain water**. Time was when Coors had mystical properties for folks in those states where it was not readily available.

32. **Beer storage**. The author recalls being given a bottle of such a strong beer, Thomas Hardy's Ale, on the occasion of the birth of his eldest child, Peter, in August 1980. The label advises keeping the beer for 25 years before drinking. It seemed inappropriate to quaff the brew when the time came.

33. **Perception of imported beer**. In a trial conducted by Guinard at UC Davis, imported and domestic US beers were compared, named and unnamed. When the products were identified, there was a heavy preference for the imports. But when conducted "blind," there was no such preference.

34. **Perception of freshness and skunkiness**. In a trial in Sacramento, we presented a well-known US lager fresh, and after we had force-aged it to develop a pronounced cardboard character. When presented as two different

samples of this famous beer, there was no preference for one or the other. We also exposed bottles of a famous beer in green glass bottles to bright sunlight, which leads to intense skunklike character. When this beer was given to consumers alongside a sample of the beer that had not been "skunked," there was a 2:1 preference for the non-skunky beer, which still leaves a third of people preferring skunky beer. Hmm.

35. **Beer turnover**. At Bass we produced an amazing beer called Lamot Reserve. It was double-fermented in that after the primary fermentation, some extra priming sugar was added, together with a second type of yeast. The beer was packed into beautiful bottles with elegant labels and test marked in a nearby town called Swadlincote. A bizarre venue: The mining community there drank draft bitters and milds, and certainly not premium bottled lagers with European credentials. The beer bombed.

36. **Volstead Act**. Named for Andrew Volstead, Chairman of the House Judiciary Committee, who oversaw the passage of prohibition into law.

37. **Ice Beer**. In the 1980s many brewers decided that it would make economic sense to transport beer in a concentrated form and reconstitute it at the point of sale. They experimented with "freeze concentration": If you freeze beer, the first thing to come out of solution is almost pure water, that is, ice. Most of the beer components remain in solution in a concentrated form. Labatt quickly realized that it wasn't going to work for the intended purpose in Canada. But Graham Stewart (their Technical Director and a close friend of the author) and colleagues, in looking for a new angle on beer marketing, felt that *ice* was a powerful beer-linked concept for Canadian drinkers, who routinely put bottles of beer into the snow to freeze out some water and boost the "oomph" in the rest of the liquid. So was born the first North American ice beer, although the

Germans had long known about Eisbier, in which beer is partially frozen at 25°F (–4°C) after fermentation and then filtered.

38. **Heileman**. Brewery founded by Gottlieb Heileman in 1858. Sold to Stroh in 1996, who duly sold out of brewing to Pabst and Miller. The Heileman name went to Pabst. The old brewery with the largest of six packs (six huge vessels adjacent to the statue of King Gambrinus, legendary king of Flanders and patron saint of beer) went to the City Brewing Company.

39. **Dimethyl sulfide**. DMS is a flavor component of many foodstuffs, including marine products, sweet corn, and parsnips. Found in many beers, especially lager beers, opinions range from those who deplore it (e.g., Foster's), those who like a middling quantity (e.g., Carling Black Label), and those with huge levels, of which Rolling Rock stands out.

40. **UC Davis**. The campus brewing program was established in 1958 and became the first of its kind on a major American university campus. Michael Lewis became the lead instructor in 1964 and retired in the mid-nineties. The author was appointed to the Anheuser-Busch Professorship of Malting and Brewing Sciences in 1999. There are extensive brewing programs both on campus and within University Extension, the latter directed by Lewis with the author's support as teacher.

41. **Miller-Coors**. A joint venture since October 9, 2007 in the United States between SAB-Miller and Molson Coors.

42. **Industrial beer**. A reprehensible term sometimes employed by a thoughtless few in the craft sector to describe the products of the largest brewing companies. Often they allege that the major companies use cheaper ingredients, additives and preservatives, etc. The reality is very different. Products such as rice are used not

because they are cheaper, but rather because they have certain quality attributes. They are actually more complicated to use than malt. The bigger brewing companies adhere to the very highest quality standards and are just as unlikely to use process aids as are smaller companies.

43. **Driving Under the Influence.** DUI means being legally intoxicated or impaired while operating a motor vehicle. The threshold for legal intoxication is typically when a breath, blood, or urine test shows a blood alcohol content of 0.08 percent.

44. At Anheuser-Busch InBev, for instance, the brewers must breathe into an alcohol detector (and be "clean") before they are allowed to drive home.

45. **Testing for drugs.** The author has heard it said that there is a correlation between the tendency to smoke marijuana and the liking for skunky beer. This is a line of research that the author will not be pursuing.

46. **Bean counters.** Bass brought in a new collection of managers who had little empathy for brewing, but simply ran the business dispassionately and at arm's length. The author would contend that the finest breweries are run by people who have a genuine love for the product in their hearts. Such is probably the case for any business.

47. **Pabst.** American company founded in 1844 by Jacob Best and since 1889 named for Frederick Pabst. Now owned by S&P Company, based in California. They now brew brands formerly owned by the likes of Heileman, Lone Star, Rainier, Schaefer, Schlitz, and Stroh. The PBR name came from blue ribbons tied around the bottle neck from 1882 until 1916.

48. **Miller Clear Beer.** A beer released in 1993 that was "water white." The color was stripped out of the product such that when it was poured into a glass it was a clear liquid with stable foam on top. The result was

confusion by customers and a product that lasted very few weeks in the market.

49. **Beer serving temperatures**. It is a myth that English ale is served warm—it is at the temperature of the underground cellars beneath the pubs. Guidelines for beer drinking temperature are

Very cold (0-4°C/32-39°F): Pale Lager, Malt Liquor

Cold (4-7°C/39-45°F): Hefeweizen, Premium Lager, Pilsner, Belgian White, Fruit Lambics

Cool (8-12°C/45-54°F): American Pale Ale, Amber Ale, Dunkelweizen, Sweet Stout, Stout, Porter, Dunkel, Helles, Vienna, Schwarzbier, Smoked, Altbier, Tripel, Irish Ale

Cellar (12-14°C/54-57°F): Bitter, Premium Bitter, Brown Ale, India Pale Ale, English Pale Ale, Unblended Lambic, Bock, Scotch Ale, American Strong Ale, Mild

Warm (14-16°C/57-61°F): Barley Wine, Quadrupel, Imperial Stout, Imperial/Double IPA, Doppelbock

50. **Bicycles**. The author cannot ride a bike, despite there being more bikes per head of population in Davis than any other city in the US. Davis is also home to the US Bicycling Hall of Fame. Neither can the author swim.

Chapter 2

1. "Ted" Heath (1916-2005), MP for Bexley and Prime Minister from 1970 to 1974 was an avowed bachelor, yachtsman, and very "old school." He tried to rein in the

unions, but there were disasters in his strategies, notably the Industrial Relations Act that established a special court that imprisoned striking dockworkers. There were two miners' strikes during his tenure as prime minister, the latter precipitating the nightmare Three Day Week.

2. There is little doubt that the concept of the unions was historically of social benefit when viewed against some tyrannical practices of unscrupulous business owners in the industrial revolution. However, by the 1970s, the far left-wing tendencies of some union leaders took the movement in extreme directions far removed from the simple expedient of "fair play for all."

3. Since being a biochemistry undergraduate at Hull University I had shifted my dream away from being the Wolverhampton Wanderers goalkeeper to being a university lecturer. It seemed like my dream would come true in 1983 when I interviewed for a new department covering biotechnology at Imperial College in London. I found the demeanor of the interviewing panel pompous in the extreme, and I was convinced after a few minutes that I had screwed things up. So I decided to be myself (or at least who I was at that time in my life) and with a degree of bluntness that (from this distance) I cringe about, I let them have it as I saw it. After an hour or so, I was asked to go and wait in the room of Professor Brian Hartley FRS...and I waited. About an hour later Hartley (a fellow north country man) burst through the door, poured out two glasses of sherry, and pronounced how delighted he was with my up-frontness and said that he had convinced them that I was the man for the job. We celebrated and in due course I made for my Crawley home via the Tube and British Rail. "I should be delighted," I said to Diane, "but it feels all wrong." Next morning I went in to BRF and met with the Director-General, Professor Bernard Atkinson. "I need to tell you that I am leaving, to go to

Imperial College." He smiled, "Yes, you are indeed leaving. But not there. I have been negotiating your transfer for weeks: You are going to Bass in Burton-on-Trent." The relief and joy was immense. I was going to a great brewing company, I was getting out of the southeast rat-race—and I would be less than 30 miles from Wolverhampton. At the age of 31 I would be able to get my first season ticket for my beloved Wolves. Academia could—and would—wait.

4. David W. Gutzke. "Runcorn Brewery: The Unwritten History of a Corporate Disaster," *Histoire sociale/Social History* 41 (May, 2008): 215-51.

5. Transport and General Workers Union (TGWU); Amalgamated Union of Engineering Workers (AUEW).

6. One of Bass's key performance criteria was Right First Time: measures of how consistently a product could be delivered within specification. Notable among these indices was the assessment with what regularity the beer after filtration was within target for its content of carbon dioxide and absence of oxygen. At Preston Brook, our score was 20 percent—in other words, on 80 percent of the runs we were *wrong* first time. This was all to do with bad brewery design, but the company was never going to invest in correcting this when faced with such industrial strife.

7. These days known as the British Beer and Pub Association (http://www.beerandpub.com).

8. Breweries in Magor (Wales), Samlesbury (Lancashire), and Glasgow.

9. Breweries in Alton (Hampshire), Burton (Staffordshire), and Tadcaster (Yorkshire).

10. Breweries in Manchester, Tadcaster, Edinburgh, and Dunston (Durham).

11. Brewery in Northampton.

12. Home of my favorite soccer team, Wolverhampton Wanderers (www.wolves.co.uk). It looked to be an auspicious time in 1983 when I got my first season ticket to Wolverhampton Wanderers (see this chapter's endnote 3). Wolves had just been promoted back to the top flight, and I could look forward to the likes of Manchester United, Everton, Tottenham Hotspur, and the rest gracing the Molineux green. The first game of 1983-1984, a 1-1 tie with Liverpool, flattered to deceive. In a word, Wolves became everybody's whipping boys.

Week after week the team would be tanked. Worst of all, there was no money to spend to improve the talent. Bad decisions and dubious investments had led to a crisis of colossal proportions behind the scenes and little did we know it (or perhaps we did), but the club was folding even faster in the corridors of power than it was on the lush turf. In successive seasons Wolves went down to the second, third, and fourth divisions. (The equivalent in the States would be the Oakland Athletics finding themselves in Class A ball after three years of free-fall and facing the likes of Rancho Cucamonga and Stockton.) Meanwhile the ground was crumbling. What bigger challenge could there be to one's loyalty? The crowd disintegrated to a very few thousand. But faith and faithfulness are all I know, in whatever I do.

It was painful to witness for somebody who remembered the glory days when Wolves were one of the top teams in the world. Each Saturday, though, with unquestioning loyalty I jumped into my car (a company one with all costs met—Bass were extremely generous to their managers) and drove to the game. Three hours later I would be home, miserable. Lost again.

In our village (Barton-under-Needwood) lived a well-known ex-football player and later manager, Tim Ward. He had played for England as a young man, and later managed a range of senior clubs. Now he was retired, and he quickly became a good friend and benevolent uncle to Peter and Caroline, the latter newly arrived. It was he that asked if I would like to meet Stan Cullis. Imagine asking a Yankees fan if she fancied a burger and a beer with Babe Ruth, and you will get a feel for this. Cullis had been captain of Wolves and England and the manager when Wolves won three Championships and two FA Cups, as well as forging a path for English clubs in Europe with a series of celebrated floodlit friendlies against the world's top teams before the days of the European Cup.

I drove Tim to a restaurant high on a hillside in green and rolling Worcestershire, where we had lunch with this godlike entity. For once I just listened, as two legends of the game reminisced and shared their tales, only one or two of which were possibly exaggerated.

Later Tim introduced me to other immortal footballers—from Sir Stanley Matthews to Billy Wright. Tim was a quiet, understated man, who knew his football, though perhaps not his beer. "Is this one OK?" he would ask if I went to his home. It was always a super strength lager—10 percent alcohol.

Meanwhile the Wolverhampton evening paper, the *Express and Star*, printed one horror story after another as they chronicled the demise of the mighty Wanderers. The worst piece of all came late in 1985. The club had so little spare cash that they had failed to pay the bill for the delivery of milk. Imagine that, if you will. Later in the same article they said that the following Saturday's match day program would be the last, as they no longer had the wherewithal to put it together and get it printed.

Instantly a light bulb went on in my head and I grabbed for paper, envelope, and stamp, and I wrote to the club. I said that no program pretty much would mean no club and that I had done a lot of writing (an exaggeration—it was scientific articles pretty much entirely) and I knew a fair bit about the club (not an embellishment—once upon a time I used to have total recall of fourth teams from years before). Could I help?

Two days later there was a reply from then chief executive, Gordon Dimbleby, inviting me to come on down, and so, later that afternoon and for the first time in my life, I found myself in the hallowed (spiritually if not physically) halls of my life's heroes. Before the week was out my first article had appeared in the program—and with what irony was the opposition that day the team from my boyhood home, Wigan.

I instantly became a regular correspondent. Each home game I would write something irreverent (some might say irrelevant), but I was in some form of heaven. And when people worked out that I was Mr. "Sidelines" and said they enjoyed what I wrote, then I puffed up bigger than any parakeet.

Wolves, though, continued to be in turmoil, and that extended to the program. At the season's close the program was franchised out—and I went with it. The new editor was Don Stanton, who I instantly gelled with. He had been a referee in senior nonleague circles, and ran the line in the Football League. He had been linesman in the 1974 FA Cup final. Don encouraged me not only to do my tongue-in-cheek column, but also to conduct a series of interviews with characters relevant to Wolverhampton Wanderers. These extended to the manager and his coaches, the Chairman and Vice-Chairman, through countless others, including the guy who made the sweet half-time tea and who scrubbed the baths.

One half-time, Don came up to me in my usual place in the stands, asking me to come with him to meet somebody who had expressed a wish to meet and who had granted an interview. Dutifully I trotted along, to be met by a tall, rugged-looking guy with long curly blond hair. It was Robert Plant, singer for the legendary rock band Led Zeppelin, also a huge fan of Wolverhampton Wanderers.

"Charlie Bamforth," he said, "delighted to meet you. I love what you write."

I mumbled that I was quite impressed with his talents, too, and that I would really like to do an interview with him. He said that it would have to wait as he was going on a world tour. "That's OK," said I, "so am I!" Which was true: I was off to a brewing conference in Australia.

Several weeks later the interview was in the "can," but not before Robert had been told by one of the elderly women in the club shop, behind which we were to do the interview, that he needed a haircut.

The only folks off-limits to me for my interviews were the existing playing staff, for reasons I never quite understood. The joke at the time was that Charlie Bamforth had interviewed lots of players, but most of them were dead. It is true that I interviewed a substantial number of former stars, not only from Wolves. Their names would read like panoply of legends to any English football supporter: Sir Tom Finney, Dennis Tueart, Joe Fagan, Joe Mercer, Billy Wright, and many more. Finney is held by many to be one of the best half dozen footballers of all time. He was born and raised 15 miles from me, and when I bought him lunch at the *Tickled Trout* in his native Preston, we were treated like royalty. And what a modest man.

Some of these articles appeared in national soccer magazines. By now, too, I was writing for all the other match day programs that Don Stanton published, namely Birmingham City, Walsall, and Shrewsbury Town. Every week I had at least two articles to write. And so, to this day, I can look to more than two to three times more articles written on soccer than on beer.

Luckily for me, Bass as a company was very much into football. And one thing that I hadn't clocked was that they had had a major involvement in Wolverhampton Wanderers.

The man who told me this was Otto Charles Darby, Chairman of Bass Brewers before Robin Manners. As a young sprog, newly into the company, I was seated next to Darby at a major company dinner. The policy was clearly one of seeing whether the youngster could hold his own in exalted company. I got off to a good start by saying something like "I wonder, Mr. Darby, would you be the same 'O. C. Darby' that features as an author on a paper about β-glucans in the *Journal of the Institute of Brewing* from the early fifties?" He was—and I could see that he was impressed. The conversation flowed, and then we got to talk about Wolves.

"You know, Charlie, as the club were restructuring when the receivers were in, they asked me to be chairman."

My pupils clearly dilated. He looked at me, plainly.

"Would you have taken that job?" he asked.

"I certainly would," was my retort. His look said it all.

"But then again, Mr. Darby, I am just a simple scientist."

"Yes, Charles," he replied. "I think you're probably right."

One of the most valuable pieces of silverware in the English football trophy cabinet is the Bass Charity Vase, which is competed for preseason by clubs in the Midland area. I soon found myself writing large chunks of the program for that event, as well as helping out by collecting money at the turnstiles. Such a role helped me truly see people for what they are worth. Some football types—notably club managers—would breeze past with their noses in the air, far too important to pass the time of day with "ordinary people." Sometimes, though, you would encounter a nugget. For instance, I well remember serving a long line of folks trying to get in before kick-off and right at the back, patiently waiting his turn, was Graham Taylor, at the time the manager of Aston

Villa. Eventually he found his way to the front and proffered a £20 bill. (He could legitimately have got in for
free, but he insisted on paying.) Admission was £2, so I
started to count out the change. "No, son, it's for charity," and he walked in. Later he would become the manager of the English national team and was pilloried and
abused. If only the millions who decried him realized
the humility of the man.

Bass also went into soccer sponsorship big time. Our
biggest brand—and the best-selling beer in England—
was Carling Black Label, and that name was used to
sponsor the Football League Championship. Its sister
brand in Scotland, Tennent's, was used to sponsor com-
petitions north of Hadrian's Wall—but was also used to
adorn the Charity Shield, a game played at Wembley
preseason between the previous season's League Champions and FA Cup Winners.

Knowing of my other life, the powers that be at Bass
decided that I should be their correspondent at Wembley. Perhaps what followed was the pinnacle of my
soccer-writing career (though I did visit the Wembley
press box on two further occasions) for I found myself
lunching with the doyens of football scribes from Fleet
Street before the big game. Nervous in such exalted
company I bolted my food (Diane would say that I
therefore did not change the habit of a lifetime) and
found my way up to the press box early. I plonked
myself in the allotted place in front of a television mon-
itor and a telephone. A schoolboy international game
was in progress, so I started to write. There was nobody
else there. One by one, the professional writers arrived,
more than one asking me for the lowdown on the warm-
up game that they had missed while enjoying their
Chardonnays and Liebfraumilchs.

Back in Wolverhampton, Wolves were doing much better. They narrowly missed going back up to the third
division, in a game where (with Diane's blessing) I had

sponsored the match ball. The following year they did go up, and then again were promoted the following year. Rapidly the fair weather fans reappeared.

And week after week I would find myself sitting alone (and later with Peter, now thoroughly indoctrinated into the religion of Wolves), desperate for success for the team. If they scored early, then for the rest of the game I would be on tenterhooks, praying for the final whistle. I could see only one team, the one in old gold shirts. I wasn't watching a football game, I was single-mindedly focused on us winning a war, with no sense of the beauty of the challenge and oblivious to the truth—that it takes two to tango. If only I had known then that this was an ultimate form of yin and yang.

One Saturday, when Wolves were not on show, I took Peter along to a minor league game in Northwich, Cheshire. I had no "ownership" of either side and the crowd was smaller and altogether gentler. There were those at Wolverhampton who seethed with rage, a myopic devotion that could spill into violence. (I had been a victim of it way back in 1968, when a Tottenham fan decided he wanted my gold and black scarf, and kneed me in the "wedding tackle" in order to purloin it.) But here in Northwich was bonhomie, softness, a higher ethos. Certainly, the on-field quality was second or third rate, but I realized that the experience I was yearning for was holistic, and that the product on the pitch was by no means the first priority.

I wrote a report on the game that dwelled as much on what happened off the field as what occurred on it, and the nations' leading nonleague football magazine published it. And there started my second soccer-writing career, as I toured nonleague grounds, reporting for this magazine and others as much on the quality of the pies and the Bovril as I did on the tactics and the score line. I turned up unannounced and was even referred to after a game on the bleak Essex coast as "nonleague football's very own Egon Ronay."

Perhaps it is only now that I realize just how accommodating Diane was with all this. By day I would be working long, long hours in a job that would increasingly take me to far-flung corners of the globe, and then on Saturdays she would wave good-bye as Peter and I headed off to our selected game. Her patience was about to be challenged even more.

I remember the moment the idea struck. I was up a ladder painting our home in Helsby, Cheshire, when my mind started thinking laterally. It went something like: Didn't make it as Wolves' goalkeeper—did make it as a writer. Hey, let's link the two—write a book about Wolverhampton Wanderers keepers.

I was down the ladder in an instant, jabbering away about my notion to Diane. She smiled: "Do it." Has any man ever had a more supportive wife?

And so over the course of the next several months I wrote the book *In Keeping with the Wolves* that was published by Don Stanton. I traced as many goalkeepers as I could who had played for Wolves (including the reserve teams) since the Second World War. Many granted interviews. I found myself on the banks of Loch Lomond and in the beauty of the Swansea bay. I chatted with television stars such as Bob Wilson. I met some in their homes, like my boyhood hero Phil Parkes, and the greatest of them all, Bert Williams, who confided that one of his worst moments was being the keeper in the 1950 World Cup debacle when the mighty England were defeated by the United States (I have interviewed two other England players from the same game— neither cherished the memory).

Diane's patience snapped just the once, when I announced that I was heading off 100 miles or more one Sunday afternoon to interview Tim Flowers at Kenilworth Castle (Flowers would become the England keeper). She hurled the newly filled tea pot at me, but it mostly missed.

Although I gave up soccer writing soon after coming to the States, I rediscovered it in 2009, interviewing former players for wolvesheroes.com.

13. Acquired Belhaven in August 2005.

14. The long handles that are used to pull the beer from casks in the cellar through suction, as opposed to taps that trigger the pushing of beer out of kegs by the injection of carbon dioxide.

15. The author is gratified by the existence of Cask Marque (http://www.cask-marque.co.uk), an organization committed to ensuring the excellence of traditional cask ales.

16. **Sparklers**. Nozzles on beer taps that ensure foam production when beer is forced through them.

17. **Diageo**. Guinness is part of Diageo, a global alcoholic beverages company.

18. How enlightened were those days for a brewing company to have two research departments: The vast majority these days have none.

19. **Cricket**. When playing cricket, it was always somewhat intimidating to turn up at the Park Royal ground of Guinness. In the changing rooms there were photographs from games in Guinness's history, such as "Guinness versus West Indies." I got my highest ever batting score against Guinness. I think it was 26.

20. **Greene King** introduced a canned version of its wonderful Abbot Ale with a widget. The customers didn't like it, so the company withdrew the device and pronounced on the can that it was "widget-free ale." Bass Ale in a can never did have a widget, the label proclaiming "For Head, Pour Quickly."

21. **Casks** are containers for traditional ales that are nonpasteurized and that rely on residual yeast to convert priming sugars to deliver low levels of carbonation. The beer is drawn from the cask either by gravity through a tap (cask behind the bar) or by drawing through pipe

runs using beer handles. Kegs are containers holding draft beer that is filtered and carbonated in the brewery and then dispensed by using gas from cylinders of carbon dioxide or "mixed gas" (carbon dioxide and nitrogen).

22. http://www.spinprofiles.org/index.php/All-Party_ Parliamentary_Beer_Group.

23. Tax rates in Europe, April 2009 (pence per pint):

Finland:	61.74
Ireland:	51.99
UK:	45.89
Sweden:	43.15
Denmark:	19.42
Slovenia:	17.95
Netherlands:	17.08
Italy:	14.76
Estonia:	12.88
Austria:	12.56
Cyprus:	12.51
Hungary:	11.12
Belgium:	10.74
Slovakia:	10.36
Poland:	9.46
Portugal:	9.05
Greece:	8.54
France:	6.91
Lithuania:	6.44
Spain:	5.71
Czech:	5.55
Latvia:	5.35
Luxembourg:	4.98
Germany:	4.94
Bulgaria:	4.82
Malta:	4.71
Romania:	4.15

24. Total revenue from taxing beer in 2004 (Euro):

Austria	202,000,000
Belgium	196,760,000
Finland	415,000,000
France	291 841 950
Germany	787,408
Ireland	457,100,000
Italy	306,714
Lithuania	31,804,000
Netherlands	324.000.000
Portugal	84,000,000
Switzerland	68,393,951
United Kingdom	4,516,000,000

Chapter 3

1. The legendary technical guy from Bass is said to have surreptitiously wiped his handkerchief across the rim of an open fermenter in Prague and then brought the prized conquest home and asked his microbiologist to isolate what he could from the handkerchief. It is supposed that the rag had not been used for anything other than stealing the yeast.

2. I was in my office at Preston Brook when one of the brewers stomped in and plunked a glass of beer down on my desk. It was decidedly blue-tinged. "Come on, good doctor, what's wrong with this?" I racked my brains but could think of nothing. "Well," said the brewer, gulping it down in one, "it tasted just fine." I had a phone call a few hours later to tell me that Nick had had his stomach pumped: The blue color was a result of a miscalculation in the addition of copper sulfate, which was added in very small quantities to snaffle

undesirable eggy aromas. But 10,000 times too much had been added—at those levels copper is poisonous. We decided that henceforth this would be one addition we could live without: better to have farty-flavored beer than kill anyone.

3. As alcohol content of a drink increases, there is a greater tendency of volatile substances to remain in the liquid. Beers being generally of lower alcohol content than wines means there is a greater delivery of aroma substances into the nose from the former. And considering there are probably twice as many flavor-active components in beer than there are in wine, then it will be realized that beer flavor is substantially more complex than is that of wine.

4. **Foam stabilizer**. Propylene glycol alginate (PGA). Seldom used in the US.

5. These were spore-forming Bacillus bacteria, present in the water supply. They would not grow in the beer and therefore spoil it, but they were present in a hibernating form (spores) and were thus of concern to the Saudis.

6. Brewers and brewing scientists love a good meeting. We have them all over the world, and I have been privileged to have been to some wonderful places to give papers: South Africa, Australia, New Zealand, Thailand, Vietnam, China, Japan, India, Singapore, Brazil, and of course, numerous venues in Europe and North America. Truth is that I haven't taken as much advantage of these trips as I should have. Diane is driven to distraction: There I am in a place like Cape Town, and rather than make my way up Table Mountain, I'm to be found in yet another hotel room, writing a paper or catching up with emails. But the brewing industry has been good to me. I have roamed the globe talking about the science of a wonderful product and have met amazing people.

A good proportion of my life has been spent in airplanes—and, until gaining a degree of enlightenment, that has been the most nerve-wracking thing I have done (save batting in cricket, that is). I am the sort of guy who likes to get to airports early, real early. The most delicious words in the world to me are "you're all set, sir" as I am handed my boarding pass, aisle seat, of course. My biggest buddy in the brewing industry is Graham Stewart, erstwhile scientific guru of Labatt and until his retirement, Director of the International Center for Brewing and Distilling at Heriot-Watt University in Edinburgh, Scotland. Graham and I have taught beer and brewing to people all over the world, most notably in Australia, New Zealand, China, and India. Graham will regale you with various stories about yours truly, such as the time I ventured out from a meeting in Shanghai to have a Chinese massage (a stocky little masseur as I recall), part of which consisted of the practitioner flaming the necks of six glass goblets and sticking them on my back. When he removed them 20 minutes later, there were six perfect red circles—where the evil spirits had departed! It took weeks for them to disappear. However, the story Graham most likes to tell is all to do with my paranoia about getting to airports early. As he tells it, "We got to London airport three hours before the plane was scheduled to take off. And I am not talking about London, England; I am talking about London, Ontario! There only was one bloody plane!" It's true. And we drank so many beers while we waited that, after I took off, Graham had to go home in a cab and leave his car at the airport. It cost him a fortune in parking!

Not only do I get to airports early, but I also insist on really, really long connection times. The reality is, though, that I once looked death in the face on an airplane. I had reconciled myself to die. I was flying from Valencia to Amsterdam in 1991. The reason I was on that particular flight rather than one directly to London escapes me. I do recall with some amusement from this distance, though, that the airplane was a Fokker. Indeed. As usual I was sitting working, with an empty seat alongside. That vacant seat meant that I had positioned myself by the window. Suddenly an old woman was pushed into the seat and an attendant strapped her in. An announcement was made. "Please listen very carefully. We are obliged to make an emergency landing in Paris. We cannot tell you whether this will be in five minutes or fifteen, but it will be soon. Please do not attempt to retrieve anything from the overhead compartments. If you wear spectacles, take them off. If you have any false teeth, take them out. When you are told to do so, adopt the brace position." There was more, but I don't recall it. They certainly didn't tell us why we were coming down. All I did know was that there was a smell of burning and the attendants looked terrified. Can you imagine what goes through your mind in such a situation? I instantly told myself that this was it, the end of my life in this realm. Amazingly, I felt more serenity than fear. There was not a single tear, no panic. I quietly thought about my family. I simply and, with composure, stared out of the window. In truth, I do not recall praying. The ground was getting nearer and nearer, and all I knew was that this was certainly not Paris. All I could see was agricultural land. It seemed that we were going to hit the ground in the middle of nowhere. The plane was not plummeting; it was descending as it would for a normal landing. Except there was that smell of burning. At last we were told to "Brace!!!!" I did, fervently. With a bang we hit the ground, and I heard the reverse thrust of the engines with a mighty roar. But how could this be? I jerked my head up and looked out of the window. We were on a runway! It transpired that a fire in the cockpit had burnt out a bunch of navigation equipment, and the pilots had been uncertain where we were. Somehow—and I still don't know how—they found their way to a small airport at Metz. I know that higher powers were with me that day, guiding that airplane down but also giving me hidden strength and serenity. It was only when I got into the small terminal that I was hit by the magnitude of what I had been through. They opened the bar and I drank a large whisky. The rest of the passengers were all Dutch—and their exuberance (or was it delayed shock?) kicked in as they enjoyed the hospitality. Then a bus arrived, and we were driven to Charles de Gaulle airport in Paris, but not before the Dutchmen had demanded that the bus stop at a liquor store so we could get fresh supplies of booze! In Paris I was escorted to a Heathrow-bound flight held back for me and was ushered into a seat. It was the proverbial getting back on the bike after falling off. From that flight on, I have always been an attentive listener to the safety demonstrations.

It was not the first time I had cheated death. The first occasions were as a small boy. I was only three and riding my tricycle when I dislodged a paving slab that a builder working on our house had left precariously propped against a wall. It toppled over and landed on my leg, crushing it. Had it landed higher up, it would have killed me. Months later, the broken leg having healed, I dashed ahead of my mother and grandmother on College Road in Up Holland, and straight in front of a car that was pulling out of the gas station. I could only have been three or four, and I can still see the man holding his head in horror, having executed his emergency stop. And I guess I can still feel the repeated slaps on my leg, newly healed or not, from my mother, administered I guess as a knee-jerk demonstration of her relief.

The scariest event of all occurred in 1977. Diane was making her first long journey driving a car, being the only driver in our family at the time. We pelted away from Sheffield along the M1, M62, M6, and M56 motorways heading for my mother's home in Helsby.

We then exited onto a side road. Diane did not adjust, and I didn't warn her quickly enough, but up ahead was a ninety-degree bend in the road, right by a canal and underneath a railway viaduct. Around the other side of the bend were traffic lights that regulate access to a swing bridge. Suddenly I yelled to "slow down," but too late. As she roared round the corner on the wrong side of the undivided road, she swerved into a high bank and the car flipped over onto its side, driver side closest to the road, and slid along with what must have been horrendous scraping. There was the crackling of sparks. But in my head all was quiet. I do remember reaching across as it happened and cradling Diane's head lest it touched the glass. Though badly shaken, we both climbed out without a scratch. And the remarkable thing? Those traffic lights were on red, stopping traffic from coming in the opposite direction towards us.

The last occasion came on June 4, 1994. I remember the date for one very good reason: It was the day that Emily arrived on this mortal coil. I had spent the night with Diane in the hospital in Chichester where our baby was born. Early that Saturday morning, weary but ecstatic, I left my wife and the precious new bundle to sleep while I drove back through the West Sussex lanes to our home. I wound through the one-way meandering streets of Petworth, behind somebody driving far too slowly for my liking. As you leave Petworth, the undivided A272 takes a right-hand turn, and thereafter assumes a long straight stretch, on which I knew I would be able to overtake. Sure enough the road was clear ahead, so I drew out to pass the straggler. At which precise moment somebody pulled out of a farm track a little further along and appeared in the oncoming lane. On seeing me careering towards him he must simply have stopped (he hadn't started to accelerate), but I was belting along at 45 or 50 miles an hour. The screech of the brakes as I slammed my foot to the ground must have been horrendous. The other car was looming nearer and nearer, but eventually my car came to a stop. I could distinctly see the horrified look of the man behind the wheel of the car in front of me. Our vehicles could only have been inches apart. I thought back to how carefully I had driven after Peter had been born 14 years earlier. What possessed me? Emily very nearly could have been, like me, fatherless at birth.

7. Some Asians lack an aldehyde dehydrogenase enzyme that is important in the metabolism of alcohol. As a result they display alcohol flush reaction, which is colloquially known as Asian Flush or Asian Glow.

8. **Kiu.** A rice-based alcoholic beverage.

9. There are now several university programs focusing on brewing in China.

10. World's top ten beer brands:

    1. Snow
    2. Bud Light
    3. Budweiser
    4. Skol
    5. Corona
    6. Brahma
    7. Heineken
    8. Tsingtao
    9. Yanjing
    10. Coors

11. One of the hazards of being a guest speaker at a Chinese brewing event is that you become the focus for the ritual of "gan bei." Everyone will come up to you and raise their glass pronouncing "gan bei!" You are expected to reciprocate and then both of you drink the entire contents of your respective glasses. It's okay for them—they get to do it perhaps once or at most a few times. But being the center of attention, you have to do it with everyone.

12. I recall well my first trip to India. I meticulously avoided drinking anything other than bottled water. I avoided anything that had been in contact with the tap water—for example, sliced fruit and greens. So when I was sitting in my seat on Lufthansa, homeward bound, I breathed a sigh of relief and tucked into the salad. Mistake, for where had that dish been prepared? The impact was a quite spectacular weight-loss regime over several days. I recall soon afterwards speaking at a dinner in London at which was present my big buddy, Graham Stewart, who led our India inward-missions. As I said in my speech, "with friends like Graham, who needs enemas?"

13. **Diacetyl**. Substance that smells of popcorn/butterscotch and which is produced by yeast during fermentation. However, prolonged contact of beer and yeast leads to the latter removing the diacetyl, and it is critical that the brewer waits for this to occur.

14. **Gushing**. The spontaneous foaming of beer when a bottle or can is opened, leading to a surge of beer out of the container. Caused by poorly soluble or insoluble materials that can populate beer.

15. **Country liquor**. A distilled alcoholic beverage made from locally available cheap raw material (sugar cane, rice, palm, coconut, etc.) with an alcohol content between 25 percent and 45 percent.

16. The most famous beer-drinking celebration is of course the Oktoberfest. This traditionally kicks off in the third weekend in September, ending on the first Sunday of October. Its history stems back to October 12, 1810, and the wedding of Crown Prince Ludwig to Princess Therese of Saxony-Hildburghausen. All the good folk of Munich were invited to the celebration at which copious quantities of beer were joyfully consumed.

The story is told of a small-of-stature brewer from Bass who joined a company delegation in the 1970s to a technical event in Germany that coincided with

Oktoberfest. Just an hour or two after stepping off the airplane, they were at the great event, quaffing liters of Marzen and devouring sides of pork, dumplings, and red cabbage. In a state of some contentment (as it were) our hero was led back to his hotel room, where he encountered for the first time in his life a duvet. Not knowing what to do, he unbuttoned it and climbed inside, refastening matters behind him. Next morning a strapping chamber maid entered the room, flung open the windows, grabbed the duvet with both mighty paws and swung. Imagine everyone's surprise when a stark naked Englishman shot out of the end of the duvet.

17. Of course an authentic hefeweissen strictly has no slice of lemon in the top. Jay Prahl, esteemed brewmaster of the Sudwerk Brewery in Davis California, talks of "NFL": "no f***ing lemon."

Chapter 4

1. **Electron spin resonance**. A form of spectroscopy that detects so-called free radicals. Back in 1984, it was Roy Parsons and I that were the first to draw attention to oxygen radicals as a key causative factor in beer staling—I presented the paper in St. Louis, Missouri, at the first Brewing Congress of the Americas. It was my first-ever visit to the United States. My wife still reminds me that I flew out just two days after Caroline was born—and she was a few days early!

2. The production volume out of the Chico brewery now exceeds 800,000 barrels.

3. **Biological oxygen demand**. A measure of the oxygen used by microorganisms to decompose waste. The more waste, the more oxygen will be consumed, and this can be used to quantify the cost of treating the waste.

4. **Anaerobic/aerobic water treatment**. In aerobic systems waste is degraded by oxygen-using bacteria, which convert the effluent to carbon dioxide and water. Anaerobic digesters use bacteria that do not use oxygen, and they convert the effluent into various end-products, most notably methane, which can be burnt to generate energy.

5. Arnold Schwarzenegger was in Chico one time to present Ken with an award. The nurses were somewhat irritated with the governor at the time and were picketing outside the brewery. Apparently this did not stop them sending in for beer! In his speech, Schwarzenegger said something along the lines that coming from Austria, he knew about good beer but then, taking a long pull from a glass of Pale Ale, pronounced the beer "grrreeatt."

6. Too many people in the craft sector feel it within their rights to rubbish the beers from the bigger breweries, calling it "industrial beer" or "fizzy yellow liquid." You will not find Ken Grossman among them. He freely admits that when he was starting his career, the folks at places like Anheuser-Busch were only too willing to help him with advice and guidance.

7. **Sonoma**. Dare I suggest that Sonoma is a more beautiful and less pretentious valley than certain others in California?

8. **Sir Francis Drake**. 1540-1596. Sea captain and politician, close to Queen Elizabeth I (who herself had a not inconsiderable love for ale). Albion, from the Greek word for *white*, historically refers to Great Britain, notably the legendary White Cliffs of Dover.

9. **Etudes sur la Bier.** Written by Pasteur in 1871.

10. Fritz Maytag is a genuinely fascinating and, in turn, fascinated man. He loves history. One of his greatest joys was the time he consulted with great scholars of Ancient Sumeria to re-create a beer based on the Hymn to Ninkasi, which you can read more about in my book

Grape versus Grain. Fritz gathered together bakers and brewers alike to make a beer as closely as possible to the ancient approaches from thousands of years before, and then presented the beer for drinking in the old way through straws at a brewers' dinner in San Francisco. Tears appear in his eyes when he tells the tale.

11. **The King is dead; long live the king**. Or rather *Le Roi est mort. Vive le Roi!*, first declared in 1422 on the accession of the French King Charles VII after the death of his father Charles VI.

12. **Weihenstephan**. In Freising, near Munich, is one of two great brewing schools in Germany, the other being in Berlin. There are not many brewing programs in the world, but other notable ones are in Belgium (both in the Flemish Leuven and Walloon Louvain-la-Neuve), UK (Heriot-Watt in Edinburgh and latterly Notting-ham), Ballarat in Australia, and, of course, UC Davis and Oregon State in the US.

13. **Reinheitsgebot**. This was decreed in 1516 by Duke Wilhelm IV in the Bavarian city of Ingolstadt in 1516, the law stating "In all cities, markets, and in the country, the only ingredients used for the brewing of beer must be Barley, Hops and Water. Whosoever knowingly dis-regards or transgresses upon this ordinance shall be punished by the Court authorities' confiscating such barrels of beer, without fail." Of course, they had no idea about yeast in those days. The original intent was twofold: to ensure that wheat and rye were available for the making of bread, avoiding price wars between brewers and bakers, but also to prevent dubious, even noxious materials entering into the recipe. The legisla-tion expanded throughout Germany—indeed, Bavaria insisted on the Reinheitsgebot being throughout the new country if they were to be unified with Germany under Otto von Bismarck in 1871. The European Court of Justice repealed the Reinheitsgebot in May 1987, basically saying that whatever was allowed in other

foods must also be allowed in beer. The Provisional German Beer Law (Vorläufiges Deutsches Biergesetz) of 1993 allows materials proscribed by the Reinheitsgebot, such as wheat malt and sugar. The Reinheitsgebot is retained as a badge of honor by many brewers who claim that the "purity law" somehow makes the beers superior. This is not necessarily the case.

14. Dan Gordon encountered a professor in Munich whose specialty was garlic. Dan was given copious quantities of garlic bulbs and proceeded to experiment.

15. The marketing strategy at the time cost my organization some money. Boston Beer wished to become members, but other members had the power of veto over which new members could be recruited. Those that perceived themselves to be the victims of the Jim Koch approach to selling beer declared that the check from Boston should not be cashed.

16. **Samuel Adams**. There is a certain irony in this name, a man who was at the forefront of Republicanism, being associated with a quintessentially British beer style, India Pale Ale.

17. It was on October 14, 1978, that President Jimmy Carter signed H.R. 1337 (curiously named "To provide for a feasibility study of alternatives to augment the water supplies of the Central Oklahoma Master Conservancy District and cities served by the District"), which contained an amendment sponsored by Senator Alan Cranston to create an exemption from taxation for beer brewed in the home for personal or family use. It became effective on February 1, 1979.

Chapter 5

1. I was there to talk to a company about egg white, which we were exploring at the time for its potential to boost the foam of beer and as a means of putting a head on novel alcoholic beverages. (I was known as "Scrambled Eggs" by my friends in Glasgow.) One of our concerns was the potential allergenicity of the albumin—a lot of people in the world are sensitive to it. I went down to Harley Street to talk with one expert. My boss had an extra suggestion: "Peter C. at Runcorn is allergic to egg white. So we'll give him some beer with it in. If he survives, we will use it. If he dies, well, we'll have a memorial brew!" He chuckled but, nevertheless, we did indeed do that very thing. Peter is alive and kicking to this day, but the egg white never did find its way into our beer. Too expensive.

2. There are three major societies worldwide for professional brewers: the Master Brewers Association of the Americas (www.mbaa.com), the Institute of Brewing and Distilling (www.ibd.org.uk), and the American Society of Brewing Chemists (www.asbcnet.org).

3. Strange: I can handle (even eat) raw fish. I adore pickled herring. And I am no fussy eater. I've eaten kangaroo, reindeer, sheeps' testicles—and fugu, the Japanese fish that has to be prepared with skill, otherwise you will die from the toxin found in various tissues.

4. The baked beans must be Heinz. They are very different from the canned beans in the US, which are altogether too molasses-like to my taste.

5. **Cornish pasties**. Comprise a pastry case within which is finely chopped skirt steak, onions, potatoes, and swede. Heavily seasoned. And the pastry is crimped—originally a device for the tin miners to hold onto while they ate. My lovely wife is Cornish, but it is not true that I married her for a lifetime supply of pasties (inappropriately, it rhymes with "nasties"). Wherever the skilled

tin miners went worldwide, you will find that the pasty culture went with them, so we are fortunate indeed here in Davis, California, to be close to gold country.

6. **Lob Scouse**. A type of lamb stew, the word deriving from the Norwegian for stew, *lapskaus*, and savored by sailors throughout Northern Europe. So, it is a meal to be found in great seaports; for example, Liverpool. Liverpudlians are known as Scousers. Oh, for my Auntie Kathleen's Lob Scouse!

7. **Ploughman's lunch**. A hearty chunk of English cheese (of which there are so many wonderful types, including from my native Lancashire), crusty bread, pickled onions, Branston pickles. The last of these is one of the foodstuffs that expat Britons pine for and that shops selling British foods in the US will stock: It originates in a village four miles south of Burton-on-Trent and is made from diverse diced vegetables pickled in a sauce made from vinegar, tomato, apple, dates, mustard, coriander, garlic, cinnamon, pepper, cloves, nutmeg, cayenne pepper, and sugar.

8. **Tripe**. Animal stomach.

9. **Dipping**. When Diane and I lived in Sheffield in the north of England, the butcher would ask us if we wanted the pork dipped. He would lower the meat into a tub of brine. The salt would help ensure that the meat retained its moist succulence during cooking—and also the rind of the meat became especially crisp and tasty.

10. **Curry**. The ultimate British food! There are more Indian restaurants than any other type in the UK. The British love affair with Indian food stems to the days of the Raj. There is a distinct westernized bias to the dishes—you won't find chicken tikka masala in Mumbai, for instance. However, there are legendary Indian restaurants in my home country—at places like Brick Lane in east London, where virtually every establishment is a subcontinent diner. This is nirvana. Mention of Brick Lane reminds me of the time that there was a

Watney's brewery thereabouts. On visiting there once I espied a tailor's shop with the name: "Manny Cohen; formerly of Savile Row."

11. See for example Bamforth, C. W. (2002) *Standards of Brewing: A Practical Approach to Consistency and Excellence.* Brewers Publications, Boulder, CO, and Bamforth, C. W. (editor, 2008) *Beer: A Quality Perspective.* Handbook of Alcoholic Beverages series, Elsevier.

12. Mention of letters after my name reminds me of one Sunday morning in England. A very casually dressed man ambled up to our door in Wisborough Green. In a very posh voice he asked for my son, who I was told had volunteered to mow his lawn when he was away on holiday. "Peter is not here right now," I grunted, "but I will write down our telephone number and you can call him." I reached for one of my business cards, thinking "this will impress him." It read "Charles W. Bamforth, B.Sc., PhD, D.Sc, C.Biol, FIBiol, F.Inst.Brew, Director of Research, Deputy Director-General." "Oh, thanks awfully," said the man, reaching into his own pocket and handing me his card. It simply said "Sir David Ratford, Her Majesty's Ambassador to Norway." My good lady later said, "I think simple cards are so much more effective, don't you?"

13. **Vicinal diketones**. There are two of these: diacetyl, with its butterscotch-popcorn aroma, and pentanedione (honey). They are produced as a side product of yeast metabolism during fermentation but are subsequently reabsorbed by the yeast, and this demands that the brewer keeps the yeast in contact with the beer until the removal is complete. Failure to do this makes for a beer that is (to my mind) far less drinkable. Diacetyl can also be made by bacteria (so-called *Pediococci*) that can live in inadequately cleaned dispense lines in bars. This will have been the problem in the bar in Portland. See also endnote 13 in Chapter 3.

14. **Helsby**. A village in Cheshire overlooking the Mersey and the Manchester ship canal, between Warrington and Chester. There is a splendid view from the top of Helsby Hill (which is owned by the National Trust), one which is only marginally spoiled by the oil refineries at Ellesmere Port. I once pointed that out to my old boss (the one who smuggled home yeast and who had the novel idea for testing the allergenicity of egg white) that by night it was quite spectacular to see the flames where they burn off the waste gases. He suggested that the view is probably something similar in Hell.

15. **Buddha in the bubbles**. I like Jean-Pierre—and not only because he referred to me as the "Pope of Foam." In fact, come to think of it, that is not a particularly impressive thing to be. When I was a young biochemist embarking on what I intended as service to humankind, I am sure I would far sooner have aspired to being the Cardinal of Cancer or the Archbishop of Atherosclerosis. But it is what it is, and speaks to rather too many papers on what puts a head on beer. And I now realize that it is merely a metaphor for the Noble Truths. Grab a cold one from the refrigerator and a glass and slip into a comfortable seat. Take a good look at the bottle: brown, green, or crystal clear; you can see the beer in there but not a hint of foam. It does not exist in this moment. It has no inherent reality. Indeed, if we are one of the millions worldwide who have the habit of drinking the beer directly from the bottle, then the foam will never exist. We all know, do we not, that when we do go the extra mile and pour the beer into the glass then the bubbles will come and they will (all things being equal) linger. The greater the vigor of our pour, the deeper the head will become. So do it now: Tip the beer over the center of the glass, cascade the nectar, and admire the whiteness as the foam surges forth. Lashings of lather. Yet, just as was the case before we lifted the bottle, there is nothing there. Not really. Look at your glass and those lovely bubbles. Chances are each

of them has a radius of something like 0.4 mm. And so if you fill the glass with foam, then there are 1.75 million bubbles right there. (Go ahead, count them if you don't believe me.) Let's say we were tiny little creatures that could walk all over and around the surface of each and every one of those bubbles. We would be confronted with an area of 35,500 cm^2. That's a little less than 40 square feet. Heck, ten bottles of beer worth of foam is tantamount to a pretty decent hotel bedroom! And still there is nothing there. Let's say that the thickness of the liquid wall around each of those bubbles is 10 millionths of a meter. That means that each of those bubbles is 95 percent gas which, of itself, we cannot see. Oh sure, we know that there is carbon dioxide in there, but to our eyes and minds there is nothing but space. And still there is nothing there, even though we are told that the bubble walls are coated in protein, which allows them to linger and survive. It has been estimated that there are 40 billion protein molecules coating every bubble. That is a thousand times more protein molecules as there are people living in California. On every bubble. And can we see them? No. So, still there is nothing there—even though each and every one of those protein molecules comprises thousands of atoms. And each and every atom has its nucleus circumnavigated by electrons. In fact, the nucleus occupies only a tiny space within the overall atom. In an atom of hydrogen, the nucleus (a single proton) has a diameter of approximately 10^{-15} m (that is one thousand million millionths of a meter), whereas the diameter of the atom itself is 10^{-10} m, or 100,000 times wider. A hydrogen atom has a solitary electron, with a diameter of less than 10^{-18} m. How tiny is that in an extremely minuscule scenario! There it is, whizzing around in a planetary orbit about a nucleus that is (relatively speaking) a distant speck. Take a moment to think about it: This solitary hydrogen atom is, to all intents and purposes, just emptiness, with only a fraction occupied by anything "solid." Work it out for yourself: The volume of the hydrogen atom is 5 x 10^{-31} m³. The volume of the nucleus is 4 x 10^{-45} m³. In other words, the volume of the hydrogen nucleus is vastly less than a millionth of a millionth of the total volume of the atom. And the electron is even more insignificant. Although atoms of other elements have many more electrons and, in turn, bigger nuclei, packed with more and more protons and neutrons, they too are merely space. Even the biggest atom is, for the most part, nothingness. And if we apply the modern-day atomic physics, which says that particles are merely waves of energy and have no particle nature at all, then this emptiness is even more complete. So all those atoms in all those proteins make for nothing more than emptiness. There is nothing there. And even if we insist, despite all this, that the foam is right there, in our eyes, an indisputable reality, then let us dissuade ourselves from any notion that even if we weren't to touch a drop of the liquid, the foam would be there for all time. It will sooner or later collapse, returning whence it came to the body of the beer. Easy come, easy go.

16. The Weights and Measures (Beer and Cider) Bill (http://www.publications.parliament.uk/pa/cm199798/cmbills/cmbills/026/1997026.htm) of 1997 declares in section 43: "Any reference to a quantity of beer or cider in any provision made under section 22 above in relation to the sale of draught beer or cider shall be construed as a reference to that quantity disregarding the gas comprised in any foam on the beer or cider." In other words, a pint is a pint of liquid and the foam needs to be atop that. Thus the presence in the North of England of "oversized" glasses in which a line indicates where the liquid should be dispensed to, leaving 1-2 cm on top for the foam. In London, chances are the glass will literally be exactly one pint, because at most there is a solitary layer of bubbles on it.

17. **Briton**. The proper term for someone who originates from Great Britain. I cringe when someone calls me a "Brit." They would not, for example, come face to face with someone from Japan and call them... you get the point? Anyway, I prefer to say that I was born English.

18. I didn't drink them; we simply decanted.

19. You can read about these studies in (a) Bamforth, C. W. (2000) "Perceptions of beer foam," *Journal of the Institute of Brewing*, 106, 229-238; (b) Smythe, J. E., O'Mahony, M., and Bamforth, C. W. (2002) "The Impact of the Appearance of Beer on Its Perception." *Journal of the Institute of Brewing*, 108, 37-42; (c) Smythe, J. E. and Bamforth, C. W. (2003) "The path analysis method of eliminating preferred stimuli (PAMEPS) as a means to determine foam preferences for lagers in European judges based upon image assessment." *Food Quality and Preference*, 14, 567-572.

20. See Roza, J. R., Wallin, C. E., and Bamforth, C. W. (2006) "A comparison between the instrumental measurement of head retention/lacing and perceived foam quality." *Master Brewers Association of America Technical Quarterly*, 43, 173-176.

21. The equipment was loaded into a U-Haul, and I drove it back across the country, accompanied by my son Peter riding shotgun. 1,992 miles.

22. **Pom**. I love the Aussies (and the Kiwis of New Zealand) despite their insistence on calling us "poms" or "pommies." The origin is uncertain. For some, it refers to "Prisoner of Mother England" and the early westerners who settled in the colonies at the "invitation" of the courts of England and for crimes such as, well, stealing a chicken. Others believe it refers to the fact that fair-skinned Britons turn the color of a pomegranate in the baking heat of the Antipodes. Regardless of the etymology, I really don't mind the name. So much better than being called a Brit.

23. There are several other sources of haze and sediments in beer, including microbes (dead and alive), starch, other complex polysaccharides from grain, and oxalic acid—which is the same stuff that is in rhubarb and which makes your tongue furry.

24. I recall being invited onto a well-known television news program once that was anchored by a famous husband-and-wife tandem. I arrived early and spent a very convivial few minutes talking to the man about his love of beer. "And does your wife like beer," I asked. He shook his head. "No," he said, "she says it fills her up too much." "Oh," I replied, "gas." "No," he deadpanned, "it's the sausages that cause that."

Chapter 6

1. In 1604 King James I of England authorized the start of a new translation of the Bible into English. It took seven years to complete and became the standard version for English-speaking Protestants.

2. The New International Version is a translation of the Bible made by more than 100 scholars and based on the original Hebrew, Aramaic, and Greek texts. The idea originated in 1965 from a transdenominational group of international scholars meeting at Palos Heights, Illinois.

3. Authentic Trappist beer is produced by six Trappist monasteries in Belgium and one in the Netherlands (the biggest).

4. In the Jewish Purim festival, a meal called *Se`udat Purim* features the drinking of substantial quantities of wine. The Talmud (a collection of ancient Rabbinic writings) says that one should drink on Purim until one can no longer distinguish between the phrases, *arur Haman* ("Cursed is Haman") and *baruch Mordecai* ("Blessed is Mordecai").

5. The Koran, which represents the words of Allah as delivered through Muhammad, is not chronological. Thus it appears that the approach to alcohol changed. Arabs had long since enjoyed their palm liquor. Early decrees from Muhammad said that no Muslim could attend prayers in a drunken condition and, as prayers were five times daily, this was certainly a force for temperance. However, the straw that broke the camel's back (actually the lamb bone that broke a man's head) was when a drunken youth hurled such a weapon at one Hazrat Saad Ibn Abi Waqqus as he recited inciteful poetry. Subsequently Muhammad was furnished by Allah with the words from Quran 5:90. And so alcohol was outlawed.

6. http://www.dharmicnaujawaan.org.gy/?q=node/51.

7. One of my favorite baseball players was Jeff Kent. He was a marvelous hitter of a ball, and I thrilled to watch him batting alongside Barry Bonds in the order for the San Francisco Giants. Bonds was even more of a lethal weapon when he had Kent distracting the opposing pitchers. But there was a very real reason why I could identify with Kent, for he was a guy whose *raison d'être* was centered outside the sport. Sure, he was technically superb, but he professed to not being particularly interested in the game. He could take it or leave it other than as a means to pay the mortgage. For him it was far more exciting to ride fast motorbikes and talk to folk about them at his shop in Texas. Baseball is to Jeff Kent what beer is to me. There will be some reading this book who will be appalled to read that, but they shouldn't be. Beer is a vehicle for me; it is not an obsession. It is not beer that motivates me. Beer and brewing have been good to me: They have given me a decent standard of living, have taken me around the world several times, introduced me to some wonderful people. And I like to drink beer. But I am no "stamp collector" or "train spotter" when it comes to beer. You won't see me in the liquor

store, salivating as I pore over all the beers on display,
desperate to try one that I have never tried before. By
the same token, you won't find me eagerly searching out
yet another brewery to visit and clamber through. Not
long ago I was invited to say a few words at a home-
brewer's event in San Diego. The invitation implored
me to come, for I was truly a "god-like entity" in the
world of brewing. Flattering, indeed, and of course I
turned up. But the prospect of tasting beer after beer
after beer was not one that I relished, and indeed I
eschewed the opportunity to judge. Often students
pitch up at my office, eagerly clutching a beer. "I would
like you to try this." "And why should I?" I reply.
"Because I brewed it." "Which is precisely why I don't
want to drink it!" Okay, it sounds mean, and of course I
have a gentle smile on my face and I usually relent. But,
truly, it is more chore and duty than anything else. Beer
is not the be-all and end-all for me. Professionally, it is
my medium—just as a sculptor will work with clay, an
artist with paints. I work with beer as I do the thing that
fills me with joy: teach. In truth, it would not matter
what I was teaching. My joy is in the performing, the
transfer of information, the exciting of others. I guess
the equivalent for Jeff Kent was launching a ball out of
Dodger Stadium, feeling the satisfaction of those he
caters to as he trots round the bases. And then he goes
home. And so I, too, come home, both physically and
metaphorically. And very few people know what that
home truly is for me. He is the guy alone in his favorite
corner in the house. The music of Dean Evenson or
Dechen Shak-Dagsay is gently drifting through the air,
so too the waft of patchouli or sandalwood. He may be
in a yoga pose, or sitting cross-legged on his burgundy
zabuton in deep meditation, or he may be reading
Osho, or Chodron, or the Dalai Lama, or many more.
There may be a glass of whiskey or Southern Comfort to
hand. He is the guy who eschews closing dinners at con-
ferences to find a quiet restaurant, for a beer, some

good food, and his book. It may be a Poirot. It may be a soccer biography. It may be Thich Nhat Hahn. It almost certainly won't be about beer. He is a man much more at peace and comfort in his solitude than among other people. His wife knows full well that he will rapidly exit a shop if it is thronged with people—the feelings of claustrophobia driving him to where he can be alone in his thoughts. Prancing about in front of 368 students in class at UC Davis is an act, a performance. Quiet moments alone with himself, or with the very few who are close to him, is when he can shed the falseness of others' expectations and find his quiet joy. I often quip when I am in company, "I'm shy—I don't like to talk to people." Everybody falls apart. "Yeah, right!" I just smile, for I know I am being deadly serious. The reality is that I am quite content with solitude. Sometimes I think I should have been a monk.

8. **The Four Noble Truths:**
   1. Suffering exists.
   2. Suffering arises from attachment to desires.
   3. Suffering ceases when attachment to desire ceases.
   4. Freedom from suffering is possible by practicing the Eightfold Path.

The Eightfold Path:
   1. Right View
   2. Right Intention
   3. Right Speech
   4. Right Action
   5. Right Livelihood
   6. Right Effort
   7. Right Mindfulness
   8. Right Concentration

9. **Right Livelihood**. Right livelihood means that one should earn one's living in a righteous way and that wealth should be gained legally and peacefully. The Buddha mentions four specific activities that harm other beings and that one should avoid for this reason: 1. Dealing in weapons; 2. Dealing in living beings (including raising animals for slaughter, as well as slave trade and prostitution); 3. Working in meat production and butchery; and 4. Selling intoxicants and poisons, such as alcohol and drugs. Furthermore, any other occupation that would violate the principles of right speech and right action should be avoided (http://www.thebigview.com/buddhism/eightfoldpath.html).

10. A fascinating account from a westerner living in Tibet before the Chinese invasion can be found in Heinrich Harrer's *Seven Years in Tibet* (Tarcher, 1996). It is quite clear how barley is a staple of the Tibetan existence— including in its fermented form.

11. My predecessor is Michael Lewis, emeritus professor, a Welshman.

12. I recommend the work of Jon Kabat-Zinn on mindful approaches to life.

13. As this book is being completed, the author has launched a freshman seminar at UC Davis on Mindfulness and Alcohol, in which a small group of students are debating and dwelling upon issues such as advertising, religion, societal impacts, and more on the perception of alcohol and its role, for better or worse.

14. It is Dr. Jonathan Powell of the MRC Collaborative Center for Human Nutrition Research in Cambridge (http://www.mrc-hnr.cam.ac.uk/research/Research-Sections/biomineral-research.html) who has drawn particular attention to the relevance of silicon to bodily well-being, including bone health. Our work in Davis on silicon in beer can be found in Casey, T. R. and Bamforth, C. W. (2010) "Silicon in Beer and Brewing."

Journal of the Science of Food and Agriculture, 90, 784-788.

15. Fleming, A. (1975) *Alcohol: The Delightful Poison*, Delacorte Press, New York.

16. **Bitters.** Strongly alcoholic drinks containing plant extracts affording bitterness; for example, angostura.

17. As Divine highlights (Divine, R. A., Breen, T. H., Frederickson G. M., and Williams R. H. [1987] *America Past and Present*, 2nd Edition, Scott Foresman and Co., Glenview, IL), the temperance movement grew in response to very real societal problems, due not least to the fact that homemade and (later) commercial whiskey was cheaper than milk and even beer. While families were imbibing whiskey at the dining table, the consumption of spirits was three times higher per head than it is today.

18. Father of Harriet Beecher Stowe, who wrote *Uncle Tom's Cabin*.

19. The Volstead Act was named for Andrew Volstead (1860-1947), who was a Republican representative from Minnesota and who, as chair of the House Judiciary Committee, oversaw the passage of the bill that introduced national Prohibition. The bill had been conceived and drafted by Wayne Wheeler (1869-1927), an attorney and zealot within the Anti-Saloon League. Justin Steuart, once publicity chief to the latter, wrote in his Wayne Wheeler, Dry Boss: An Uncensored Biography of Wayne B. Wheeler (New York: Fleming H. Revell Company, 1928): "Wayne B. Wheeler controlled six congresses, dictated to two presidents of the United States, directed legislation in most of the States of the Union, picked the candidates for the more important elective state and federal offices, held the balance of power in both Republican and Democratic parties, distributed more patronage than any dozen other men, supervised a federal bureau from outside without official authority, and was recognized by friend and foe

alike as the most masterful and powerful single individual in the United States."

The Volstead Act read (in part):

Title 1. To provide for the enforcement of war prohibition.

The term "War Prohibition Act" used in this Act shall mean the provisions of any Act or Acts prohibiting the sale and manufacture of intoxicating liquors until the conclusion of the present war and thereafter until the termination of demobilization, the date of which shall be determined and proclaimed by the President of the United States. The words "beer, wine, or other intoxicating malt or vinous liquors" in the War Prohibition Act shall be hereafter construed to mean any such beverages which contain one-half of 1 per centum or more of alcohol by volume:...

Title 2. Prohibition of intoxicating beverages.

Sec. 3. No person shall on or after the date when the eighteenth amendment to the Constitution of the United States goes into effect, manufacture, sell, barter, transport, import, export, deliver, furnish or possess any intoxicating liquor except as authorized in this Act, and all the provisions of this Act shall be liberally construed to the end that the use of intoxicating liquor as a beverage may be prevented. Liquor for non-beverage purposes and wine for sacramental purposes may be manufactured, purchased, sold, bartered, transported, imported, exported, delivered, furnished, and possessed, but only as herein provided, and the commissioner may, upon application, issue permits therefore...

Sec. 6. No one shall manufacture, sell, purchase, transport, or prescribe any liquor without first obtaining a permit from the commissioner so to do, except that a person may, without a permit, purchase and use liquor for medicinal purposes when prescribed by a physician as herein provided, and except that any person who in the opinion of the commissioner is conducting a bona fide hospital or sanitarium engaged in the treatment of persons suffering from alcoholism, may, under such rules, regulations, and conditions as the commissioner shall prescribe, purchase and use, in accordance with the methods in use in such institution liquor, to be administered to the patients of such institution under the direction of a duly qualified physician employed by such institution.

Sec. 7. No one but a physician holding a permit to prescribe liquor shall issue any prescription for liquor. And no physician shall prescribe liquor unless after careful physical examination of the person for whose use such prescription is sought, or if such examination is found impracticable, then upon the best information obtainable, he in good faith believes that the use of such liquor as a medicine by such person is necessary and will afford relief to him from some known ailment.....

Sec. 18. It shall be unlawful to advertise, manufacture, sell, or possess for sale any utensil, contrivance, machine, preparation, compound, tablet, substance, formula direction, recipe advertised, designed, or intended for use in the unlawful manufacture of intoxicating liquor.....

Sec. 21. Any room, house, building, boat, vehicle, structure, or place where intoxicating liquor is manufactured, sold, kept, or bartered in violation of this title, and all intoxicating liquor and property kept and used in maintaining the same, is hereby declared to be a common nuisance, and any person who maintains such a common nuisance shall be guilty of a misdemeanor and upon conviction thereof shall be fined not more than $1,000 or be imprisoned for not more than one year, or both....

Sec. 25. It shall be unlawful to have or possess any liquor or property designed for the manufacture of liquor intended for use in violating this title or which has been so used, and no property rights shall exist in any such liquor or property....No search warrant shall issue to search any private dwelling occupied as such unless it is being used for the unlawful sale of intoxicating liquor, or unless it is in part used for some business purposes such as a store, shop, saloon, restaurant, hotel, or boarding house....

Sec. 29. Any person who manufactures or sells liquor in violation of this title shall for a first offense be fined not more than $1,000, or imprisoned not exceeding six months, and for a second or subsequent offense shall be fined not less than $200 nor more than $2,000 and be imprisoned not less than one month nor more than five years. Any person violating the provisions of any permit, or who makes any false record, report, or affidavit required by this title, or violates any of the provisions of this title, for which offense a special penalty is not prescribed, shall be fined for a first offense not more than $500; for a second offense not less than $100 nor more than $1,000, or be imprisoned not more than ninety days; for any subsequent offense he shall be fined not less than $500 and be imprisoned not less than three months nor more than two years....

20. So-called "3-2" beer. It is, of course, 4 percent ABV. Because 1mL of ethanol weighs 0.70g, ABV is higher than alcohol by weight (ABW).

21. Among F. D. Roosevelt's other quotations are
    I sometimes think that the saving grace of America lies in the fact that the overwhelming majority of Americans are possessed of two great qualities—a sense of humor and a sense of proportion.
    We must remember that any oppression, any injustice, any hatred, is a wedge designed to attack our civilization.

22. Anti-alcohol organizations in the US:
    Mothers Against Drunk Driving (www.madd.org)
    Robert Wood Johnson Foundation (www.rwjf.org)

Center on Addiction and Substance Abuse (www.
casacolumbia.org)

Marin Institute (www.marininstitute.org)

Center for Science in the Public Interest (www.cspinet.
org)

Center on Alcohol Marketing and Youth (http://camy.
org)

Office of Alcohol and Other Drug Abuse (www.
ama-assn.org/ama/pub/physician-resources/public-
health/promoting-healthy-lifestyles/alcohol-other-drug-
abuse.shtml)

Center for Substance Abuse Prevention (http://
prevention.samhsa.gov)

American Council on Alcohol Problems (http://www.
calcap.org/Home.asp)

23. Joseph Anthony Califano, Jr., was among other things a
special assistant to President Lyndon B. Johnson, serving as the senior domestic policy aide.

24. They refer here to an article in the *New York Times*
(http://www.nytimes.com/2009/06/16/health/16alco.
html?_r=3) which says, for example:

*For some scientists, the question will not go away. No
study, these critics say, has ever proved a causal relationship between moderate drinking and lower risk of
death—only that the two often go together. It may be
that moderate drinking is just something healthy people
tend to do, not something that makes people healthy.*

Even if it were true that there is no causal link between
alcohol consumption and health (and the journalist has
hardly presented a balanced case—see Chapter 8,
"Looks Good, Tastes Good, and..."), then at the very
least the inference is that moderate consumption of
alcohol does not *detract* from a healthy existence. The
article makes plenty of play on the source of funding for
the studies that claim benefits from drinking. It is an

obvious pitfall—the risk of being perceived as being self-serving. However, could one not equally argue that the millions devoted to attacking the alcohol industry, underpinning somewhat jaundiced and unbalanced rhetoric, are equally disingenuous?

25. And I contend that beer is very much a "foodstuff"—see, for example, Bamforth, C. W. (2004) *Beer: Health and Nutrition*, Blackwell, Oxford.

26. C. K. Robertson edited a fascinating book called *Religion and Alcohol* (Peter Lang, New York, 2004) in which, among other things, there is a reasoned critique of the extreme religious stance that wherever wine is referred to in the Bible, it is actually nonalcoholic grape juice. The book starts with reference to the Inklings, a group of authors including J. R. R. Tolkien (*Lord of The Rings*) and C. S. Lewis (*Chronicles of Narnia*), who would socialize over beer in an Oxford pub. Tolkien once said of Lewis "(he) put away three pints in a very short session we had this morning and said he was 'going short for Lent.'" Ignatius, the third Bishop of Antioch wrote: "Do not altogether abstain from wine and flesh, for these things are not to be viewed with abhorrence, since the Scripture says, You shall eat the good things of the earth...wine makes glad the heart of man, and oil exhilarates, and bread strengthens him. But all are to be used with moderation, as being the gifts of God. For who shall eat or who shall drink without Him? For if anything be beautiful, it is His; and if anything be good, it is His." I take the liberty of interpreting the word *wine* more liberally to mean alcoholic beverages.

Chapter 7

1. I was born not that long after the end of the last world war. I had a ration book as a baby—vouchers that qualified one to receive the staples of life (there was no token for beer for adults). I grew up listening to war stories and also gleaning tidbits about the heroics of those who would not willingly volunteer their stories. Blokes like my Uncle Bert. I had been told that my father, who was so much older than my mother (see endnote 15 of this chapter), had fought in the First World War. When I was rather older, someone drew my attention to the work of Fred Holcroft, a Lancashire author, who had written a short book called *Finest of All: Local Men on the Somme*, one of a series on the history of the First World War. I have no idea how he introduced a quotation from my father in this book, but there it is on page 62. The piece reads:

   J. W. "Jock" Bamforth of Birch House in Up Holland, in later life a well known and respected headmaster, had just completed his teacher training course at Saltley College, Birmingham when war broke out in 1914 and together with several fellow students he volunteered for the 1/8 Royal Warwickshire Regiment.... He was in the first wave to go over the top. "At 7.29 am we were ordered to lie on the parapet and at exactly 7.30 am we were given the order to advance. We had to advance at walking pace, rifles in the air (a new method). We had hardly got the order when every gun in Germany blazed away at us, especially machine guns which did all the damage. Before we crossed No-Man's Land we had lost practically half the Battalion. We reached the fourth German line and held it for about four hours when we had spent our stock of bombs and ammunition and we were not able to communicate with our H.Q."

   It seems that my father was one of 70 men who were almost cut off and who were forced to fight their way back. (Imagine that, for a moment—walking in and out of the gates of hell with a need to repeat the nightmare soon after.) The Corps of which Jock Bamforth was a part suffered more casualties than any other on that dreadful July 1st, 1916. All told, there were more than one million casualties on the Somme. On that first day alone 19,240 soldiers in the British army died, to make it the bloodiest day in the army's entire history. My father survived with a bullet in the arm. As I sit here, more than 90 years later, I quietly contemplate how my God carried half of my genes to safety through the mud, filth, and blood by a river that I have never seen in Northern France. And I ponder quietly the sheer unlikelihood of innumerable occurrences of survival against the odds that must have occurred throughout the convoluted path that brought my own set of chromosomes into juxtaposition.

2. Scotland is lager (and whisky) country: Tennent's lager was first brewed in Glasgow in 1885. And didn't I get to hear it! My boss—the irrepressible Tony referred to elsewhere in the endnotes, once dispatched me to the great Scottish City in the mid-1980s to tell them that their lager needed some attention if it was to match the Carling in Burton. The words "Sassenach," "wee," and "bastard" were among those fired at me before I hastened back to the airport. "We have been brewing lager beer in Scotland for a hundred years. Piss off back to your ales." Ah, tolerance.

3. The classic English style is ale, made from excellent barleys that germinated readily and consistently and therefore could be dried to quite a high temperature, and could be mashed with relatively simple technology (see Appendix A for a simple description of malting and brewing). However, the barley available to the Germans was less good and germinated unevenly. For this reason, the breakdown of the grain to render the starch accessible had to be continued in the brewhouse, and before that, the malt had to be relatively lightly kilned so as to preserve enzymes for the ensuing mashing (they did not

of course know about enzymes—all of these things were worked out experientially). The light kilning led to the inevitable low colors in the resultant beers. In the brewhouse, they worked out that they needed to start out the mashing at a relatively low temperature (we now know this was to allow the action of the more heat-sensitive enzymes) but that a higher temperature was needed to efficiently produce sugars for fermentation. They had no thermometers to measure temperature, so they resorted to taking out a proportion of the mash, boiling it, and then adding it back to the main mash to raise the temperature. The process is called "decoction" mashing. Finally, the fermentation and maturation process hit upon was slow at a relatively low temperature—the beer being stored in caves. The word "lager" means "to store."

4. As a boy, there was a huge map in the front of my classroom which marked in red the parts of the world that were within the British Empire. I think it was somewhat out of date and was presumably there to instill pride in the kids. It was John Wilson (pseudonym Christopher North) who wrote in his five-volume *Noctes Ambrosianae* (W. J. Widdleton, New York, 1867): "His Majesty's dominions, on which the sun never sets," which is now generally transcribed to "The sun never sets on the British Empire."

5. Stop press: The Dutch brewer Het Koelschip have launched 45 percent ABV Oblix. Pretty soon we will have beers that you can fuel your car with.

6. The higher the alcohol content of a beverage, the greater is the tendency of the alcohol to retain volatile materials in the beverage. For this reason, the complexity of the aroma in a regular strength beer is likely to exceed that in a typical-strength wine.

7. Once again I was in Glasgow. The year was, perhaps, 1986. As the young Research Manager of Bass, I was in town to take part in the annual road show that the headquarters technical managers were expected to make to

the various regions' breweries. We had the brewery at Wellpark, and another in Edinburgh, so one of our regular jaunts was to Scotland. Being a Bass employee, I naturally was expected to stay at the Crest Hotel downtown. It was "one of ours" (a sentiment which would become even more important a few years afterwards, when Bass sold all its breweries and bought hotels instead—lots of them—making it, through owning the likes of Holiday Inns, the biggest hotelier in the world). On that particular evening I needed some cash, so I wandered up the road to the nearest Nat West ATM machine, which happened to be in Blythswood Square. That in itself was a risky thing to do, for therein was the favored haunt of the local ladies of the street. As soon as I had withdrawn my crisp tenners, I was approached by an attractive young woman who wondered if I would like to make an investment in her services. I declined and hurried across the street, making my way briskly back towards the security of the hotel, which is when I was accosted by the little man with the bloody face. He didn't make a straight line for me, but rather meandered somewhat like a yacht cast about on a turbulent ocean. Somehow he got close to me, and lifted his hand to gently prod me in the chest. I noticed that he was holding a brown paper bag with the outline of a bottle inside. Blood was trickling from a cut somewhere near an eye. "'Scuse me, 'scuse me, Jimmy," he said. (It was like the movies, the archetypal wee Scot in his cups, referring to everybody by the diminutive form of James.) I just looked at him with a blank expression, while inwardly I guess I was girding up for confrontation, but his countenance was actually one of curiosity and puzzlement. "'Scuse me," he said once more, "where am I?" I told him and, ignoring his profuse gratitude, I sped on my way.

8. See endnote 1 in Chapter 5.

9. Two Dogs was supposedly developed in Australia to deal with a lemon glut.

10. It is always beer shown on the photo behind the television news anchor, never vodka or even wine. The reason is that the latter glass might just as well contain water or fruit juice but the foam tells the viewer that the story is about booze. So the link is drawn between beer and bad behavior, despite the fact that it is more likely shots of hard liquor that present the greater problem.

11. They are described in repulsive detail in www.college-beergames.com.

12. Cider in the UK refers to an alcoholic beverage, a.k.a. "hard cider" in the US. In the English west country, the cider is sometimes called Scrumpy.

13. **Grammar school**. When the author was a boy, all primary school children after their eleventh birthday sat an examination called the "eleven plus." If you passed, you went to a more academically demanding school, the Grammar School. If you failed, you went to the "Secondary Modern." In the 1970s the system was scrapped in favor of comprehensive education.

14. Legal drinking ages worldwide:

None: Albania, Armenia, Azerbaijan, Comoros, Equatorial Guinea, Gabon, Ghana, Guinea-Bissau, Jamaica, Kyrgyzstan, Morocco, Solomon Islands, Swaziland, Togo, Tonga, Vietnam.

Sixteen: Antigua, Barbados, Belgium, Georgia, Germany, Greece, Luxembourg, Malta, Norway, Poland, Portugal, Spain (Asturias)

Seventeen: Cyprus

Eighteen: Algeria, Argentina, Australia, Bahamas, Barbados, Belarus, Belize, Bermuda, Bolivia, Botswana, Brazil, British Virgin Islands, Bulgaria, Cameroon, Canada (19 in some provinces), Cape Verde, Central African Republic, Chile, China, Colombia, Congo, Costa Rica, Croatia, Czech Republic, Denmark, Dominican Republic, Ecuador, Egypt, El Salvador, Eritrea, Estonia, Ethiopia, Finland, France,

Guatemala, Guyana, Hungary, Indonesia, Ireland, Israel, Jamaica, Kazakhstan, Kenya, Latvia, Lesotho, Lithuania, Malawi, Mauritius, Mexico, Moldova, Mongolia, Mozambique, Namibia, New Zealand, Niger, Nigeria, Norway, Panama, Papua New Guinea, Peru, Philippines, Russia, Samoa, Seychelles, Singapore, Slovak Republic, South Africa, Spain (except Asturias), Sweden, Thailand, Trinidad and Tobago, Turkey, Turkmenistan, Uganda, Ukraine, United Kingdom, Uruguay, Vanuatu, Venezuela, Zambia, Zimbabwe

Nineteen: Nicaragua, South Korea

Twenty: Iceland, Japan, Paraguay

Twenty-one: Fiji, Pakistan (non-Muslims), Sri Lanka, United States

15. My father, John William ("Jock") Bamforth, was born on September 5, 1893. Queen Victoria was on the throne, which straightway tells you how half of my genome is pretty ancient. By all accounts, my father was one of the most decent men ever to draw breath: school headmaster, Justice of the Peace, church organist (at St. Thomas's), lay preacher. Enter Edith Halwood, spinster of this parish, born September 7, 1920. She had been appointed as the assistant at Digmoor School in Up Holland, where Jock was the headmaster. On October 17, 1949, they married; on December 10, 1950, was born unto them a son (John Richard) and then, on February 8, 1952, Jock Bamforth died, just months after being diagnosed with cancer of the pancreas, one week before the passing of King George VI. By all accounts, the latter's demise attracted more publicity, but I am reliably informed that St. Thomas's was packed to the gunnels and overflowing as my father's funeral took place.

It was Friday the 13th of June, 1952, when I elected to take my inaugural bow at the Sandbrook Nursing Home in a district of Up Holland called Tontine. I am told a thunderstorm was in full venom overhead—read into that what you will. I was Charles on account of Uncle Geoff, whose second name was Charles, and not as some daft whim from a royalist mother. And the second given name was William, after my father, and his father before him, and his father before that, and his old fella, too. And now my son is Peter William, and my grandson is Aidan William. What you might call a long line of Willies.

I often wonder what it must have been like for my mother. I have never asked: It's not the sort of conversation I could ever have comfortably with her. To be left, though, with a podgy 15-month-old while carrying another unknown quantity, while burying a husband old enough to be her own father must have been somewhat traumatic. I wonder that she stayed sane, and that I grew up "normal." I did not speak until I was three years old. Except for one word: "no." It is, of course, a very useful word, one which I could stand to recultivate, for I find myself altogether too ready these days to say "yes" and then become stretched in rather too many directions. Of course, as a toddler back in the early fifties I had no notion of how valuable a few choice words could be. It was my brother who did the speaking for me. I have no recollection of he and I having some hidden and secret language through which we communicated out of earshot of the adults. However, it seems that John did the speaking for me, such as "Charles don't want no nuninons," which, I have to say, makes me rather suspicious from this distance of time, for just about the only foodstuff (even beer) that I could not do without is onions. When I was an older child, they even comprised a meal for me in their entirety: boiled onions, drizzled with butter and accompanied by plentiful bread and butter.

Once I learned to talk, there was no stopping me. I have heard it said that there has been no shutting me up since. My wife will tell you that this is very much *not* the case. She sees the real me: a man of relatively few

words, at peace with his newspaper and not trying to
live up in his "real world" to a reputation for jocularity
and "hail fellow, well met." It wasn't until I got to UC
Davis that I encountered (or was prepared to listen to)
the honesty of those who would have me hold my
tongue. I pitched up, laptop in hand, at the Silo Pub on
campus, having been invited to give a talk at an event
where they were dishing out free beer. The students
took a vote and opted just to have the beer.

My childhood was, on balance, an exceedingly happy
one. It is remarkable but true that I never really knew
what I was missing not to have a father. It is only when I
ponder what I didn't do—like learn to ride a bike or
swim—that I reflect on some of the things that perhaps
a dad would have worked with me on. Perhaps I would
have been less than the thorough incompetent I
became in practical jobs around the house if my father
had lived. As it was, my mother was very protective.
And also, as a young widow, she needed to be careful
with money. I prayed to God for an electric train set,
like the one that my cousin John had down on the farm.
I got a clockwork one. I wanted to be a cub scout, but
the nearest I got was a well-thumbed book all about it: I
knew all the badges and how to qualify for them, but
that was as near as I got to a green uniform. I desper-
ately hoped for a drum kit, but made do with some
closed encyclopedias and two 12-inch rulers and beat
them to pieces in tribute to Dave Clark, Chris Curtis,
and the ilk. It took until 2006 for a real kit to arrive in
the Bamforth household—it's my daughter Emily's
actually, but I muscle in with a mean John Bonham
"cover"—the neighbors are very tolerant.

As a child I learned to respect possessions. It could be
taken to extremes, though: I was forbidden to sit on the
settee for fear of damaging it. "Be coming off that
eighty pound couch" was my grandmother's salutation.
She was the toughie in the family. To this day I adore
lowering myself real s-l-o-w-l-y into steaming baths that

are unhealthily hot. Back in the fifties I started this habit—and had barely got my butt onto the bottom of the bath before Nana would be hammering on the door, demanding that I "be coming out of that bathroom with all that steam ruining the paint." I have obliterated from my memory how I would be locked under the stairs in a dark closet for being naughty. I know that she loved me, but she had a mightily strange way of showing it sometimes. Her toughness had undoubtedly been forced upon her.

Jane Petty, my grandmother, was the twelfth of thirteen children, born into a family that had farmed from way back. It was inevitable that she would "court" and marry a farmer's son, and her chosen beau was 42-year-old Richard Halwood. In a period of six years there were four children, my mother being the youngest. And when little "Edi" (my mother) was only seven, Richard died. Jane Halwood stayed on as tenant at Captain Leigh's Farm in Up Holland, to be farmer and mother. She was tough. And yet so loving of the little boy who was her favorite among the grandchildren. I remember the delicious warmth of sitting on her lap as she taught me to tell the time. I think of our closeness as we traveled on the bus to Wigan, to eat pies and drink tea at Poole's café, for me to sit on a pouffé as she had her hair set by Mrs. Bannister. Oh, that smell of egg shampoo! Despite the warm and protected environment, the spirit of adventure was already taking me away from the security zone. The quarry was less than a mile away but to a three-year-old, it was another world. Over the wall I went, deep into the gloom beneath forbidding trees. All was quiet. I wonder quite what I would have done if a friendly miner hadn't peered over the wall (he had seen a podgy little chap headed in there) and, grinning, extracted me, took me back to "Kenwood," and asked of my frowning grandmother, "Is this little chap yorn?" Thank God it was him and not one of the fearsome looking nuns that promenaded up and down College

Road, or one of the priests from the renowned Roman
Catholic college that was spitting distance away. My
mother and Nana, the two adults in my home until my
grandmother died in 1964, were always very careful
with money—for instance, most of my clothes were
hand-me-downs from my brother—but I cannot con-
sciously say I wanted for anything that was genuinely
necessary. Indeed, I had some things that I was less
than enthusiastic about, like weekend trips (when I
invariably felt nauseous in the back of the car) to places
like the Yorkshire Dales, Derbyshire, and, most dreary
of all, Wales. (I guess I was projecting myself to a future
of gloom on dry Sundays, when the pubs of the princi-
pality were closed on religious grounds.)

But they were happy times. To this day my heart leaps
when I am walking along the narrow roads in Up Holland or the other villages of Billinge and Orrell where I
lived. My mind's eye can still see the miners walking
along the road, black in the face, and spitting copiously
on the flagstones. I can still see myself holding a
streamer descending from a vast banner as I walked
with all the other parishioners on Whit Sunday, and I
can still see myself poking at the bloody blisters on the
back of my heels, induced by the brand-new sandals
that had been bought for me especially for the occasion.
I can still smell the tomcats in Miss Snape's bakery
where my mother bought cakes. I can sense fat Sydney
Stevens cutting my hair, a requirement I detested then
just as much as I do now. I can see myself sitting in Mrs.
Bullough's (the dentist's) chair and the gas mask being
lowered over my face as yet another tooth was extracted
(too much Ribena—but my mother swore that this
sugar-loaded stuff, dissolved in boiling water, was just
the elixir to beef up her scrawny infant. Sadly she was
less fastidious about my teeth-cleaning habits). I can see
the red Ribble buses and the onion sellers on their bikes
and the gypsies being turned away by my grandmother
with a "be gone—what you need is a dose of bug and

louse powder." Their curses meant nothing to her. I can see myself on Sunday afternoons in the country lanes collecting nature's bounty to show-and-tell at school—rosehips and catkins and sycamore wings. And I count my blessings. Yet I will never understand why until I reached my mid-teens friends were discouraged or why I never had a birthday party.

Chapter 8

1. Beer is a significant source of some B vitamins, but not thiamine (vitamin B_1).

2. In my lab I only take undergraduates and MS students, never PhDs. There are no jobs for PhD students in the brewing industry, with the possible exception of some companies in Japan and China. What the industry is looking for is managers and brewers, not researchers. And very little in a five-year PhD program will prepare one for the night shift. My much-missed late father-in-law, Don Dunkley, would say that to have a PhD was to be "educated to the heighth (sic) of ignorance." Having said which, he was pretty proud of his son and son-in-law, who had them.

3. In the spring of 1978, when I was towards the end of my post-doctoral appointment in the Department of Microbiology at Sheffield University, an advertisement took my eye—there was a need for a microbial enzymologist. It was exactly up my street, even more so when I scanned through the entry to find that the potential employer was the Brewing Research Foundation. I traveled down to Nutfield, Surrey, for an interview on Thursday, and on Saturday a letter plopped through the letterbox of our Woodseats home to offer me the job. Two days. I thought, "Wow, they want me." And so I went into brewing.

4. Ring.

5. French president Jacques Chirac was quoted as saying (about the British) to German and Russian leaders in 2005, "One cannot trust people whose cuisine is so bad." It seems, though, that he thought Finnish food was worse. Vive la difference, say I. Tolerance, *une autre fois*.

6. One unit is 8g or 10mL of alcohol. So, if your beer is 5 percent alcohol by volume, than means that a 12-ounce serving (355 mL) is 17.75 mL, or 1.8 units. In other words, a couple of such beers each day would be the maximum.

7. Owens, J. E., Clifford, A. J., and Bamforth, C. W. (2007) "Folate in Beer." *Journal of the Institute of Brewing*, 113, 243-248.

8. Probiotics are certain bacteria that are loaded into foods (for example, yogurts) on account of their potential benefits to large intestine health by reinforcing the levels of such organisms naturally present. Prebiotics are substances that these beneficial bacteria can feed on, and so eating foods containing prebiotics is an alternative strategy for boosting the level of these "good bacteria" in the bowel.

9. Ferulic acid is also the molecule that is converted by the special ale yeast used to make hefeweissen to the substance that affords the clovelike aroma.

10. In other words, around the recommended limit referred to in endnote 6.

11. In fact, I have concerns about the interpretation of many studies (including my own) that address beer in the context of health. This is not because I doubt their veracity in any way. Rather it is because sometimes desperate and hurting people cling to the hope that they engender. Following the press release on our silica work, I received several heart-wrenching messages from people suffering from osteoporosis, people who

confused me for a medical practitioner. It was harrowing stuff, and I felt such regret that they had misinterpreted what we had really done, which was restricted to studying the factors that impact silica levels in beer.

12. **Glycemic index**. Classifies carbohydrates according to their impact on the sugar content of blood.

13. The calorie content of beer is most readily quantified as

    Calories (kcal/100g) = 6.9(A) + 4(B-C)

    Where A is alcohol (% by weight), B is real extract (% by weight), and C is ash (% by weight).

    To a first approximation, the real extract is the level of everything from the wort left behind after fermentation except for ash (inorganic minerals). For the most part, this residual material is carbohydrate and protein.

14. Charles Dickens has Mr. Micawber say: "Annual income twenty pounds, annual expenditure nineteen pounds nineteen and six, result happiness. Annual income twenty pounds, annual expenditure twenty pounds ought and six, result misery." Substitute calories for money and you get the idea.

Chapter 9

1. Which is not to say that yeast is unimportant. It most decidedly is; however, it is readily controlled if brewers pursue their storage, propagation, and yeast management strategies properly. Yeast is not subject to the same seasonal and geographic vagaries as cereals and hops.

2. As I wrote in *Brewer and Distiller International*, 5(6), 28-29:
   Would it be naïve for me to suggest that people should be regarded as an opportunity rather than as a major line item? When a company is founded on the basis of

its people we have robustness and stability. When a company per se forms the basis, then people issues become a precarious balance and instability ensues. To my mind the workforce needs to be considered as living individuals rather than as assets. They have needs, both physical and emotional. And if these wants are satisfied and an individual relates to the company for having nurtured those desires, then both the individual and the company benefit. And when we are talking people I am being all-inclusive. Every single employee of a company matters—and so I am filled with distress to read recently of one chief executive saying that only a handful of employees really make a difference. At risk of being accused of naïveté I would suggest an ideal scenario: A company celebrates the fact that its product—beer—is a unique and special component of society and embraces that fully in striving for excellence of quality, responsibly and genuinely positioning their brands from a proposition of a wholesome and fulfilling lifestyle for the consumer (people again!). The company is staffed from top to bottom by individuals with, sure, appropriate technical qualifications in their various fields, but who are also recruited on the basis of their qualities as human beings. In particular, people who not only can lead and be members of teams but who are enlightened (or enlighten-able) in matters spiritual. Relax, I am not going to get heavy on you here, but I would draw attention to companies (such as AOL Time Warner, Sony, Toshiba and Google) who have embraced tools such as meditation, recognizing the benefits this has for interpersonal relationships and understanding, as well as clearing the mind. By spiritual I am not talking religion, I am talking peaceful co-existence and core values of empathy and goodwill. And so we have a management and a workforce in which people are at the heart. They are motivated. They interact with mutual respect. They are minded to deliver the product to the best of their ability. And quality product means market share and profitability.

3. The calculated value was 3,188.8g CO_2 per six-pack. It seems that the major contributors are the cost of supplying malt and bottles to the brewery and the cost of refrigerating beer in trade.

4. A certain amount of thought needs to go into this. For example, the *Mail Tribune* in Southern Oregon in describing the issue of bikes to one pub brewery's staff said, "The bikes will reduce the restaurant's carbon footprint by getting workers out of cars; enhance workers' health and fitness; and free up parking spaces for tourist-oriented Ashland." A strange concept, to lessen local traffic to encourage traffic coming much greater distances.

5. In Japan, if a product contains less than 25 percent malt, it attracts far less taxation—so-called happoshu products that contain high levels of adjunct. If a drink is made from a grist with zero malt, then the tax is even lower. These are so-called "third way" beer products. None of these products can be called beer, but they are packaged like beers and have the same imagery. Their cheapness means that they are growing substantially in volume, at the expense of beer.

6. Heymann, H., Goldberg, J. R., Wallin, C. E., and Bamforth C. W. (2010). A "beer" made from a bland alcohol base. *Journal of the American Society of Brewing Chemists*, 68, 75-76.

Chapter 10

1. Louis Pasteur, Charles Cagniard-Latour, and Theodor Schwann were pioneering scientists studying the biology of fermentation. Dr. Ray Anderson, once of Allied Breweries, writes eloquently about the history of brewing science. His work and that of others can be read at www.breweryhistory.com.

2. Ten years later Eduard Buchner was awarded the Nobel prize for his work.

3. Marmite is the only product I know where the company responsible bases a marketing strategy on saying you may hate it (www.marmite.com—*I* do, whereas my wife loves it). It is basically yeast extract. The Aussies swear by their equivalent, Vegemite.

Conclusions

1. I *really* wanted to be a soccer goalkeeper! As a boy I had a plastic soccer ball that had burst a long time ago. It for the most part had retained its spherical shape, except for a portion that was completely flat, so the ball was thoroughly asymmetrical. Hour after hour I would be alone, throwing this ball as hard as I could against a wall. Of course there was no predicting how it would rebound. If it hit the wall with the curved surface, then it would bounce straight back into my arms. But if it hit the wall on its flat edge or at the point where the curved and collapsed parts met, then it would come back at some wacky angle—left or right, high or low. And this skinny and lonely 13- or 14-year-old lad would dive and catch it, over and over and over again. My reflexes were remarkable. And I had no fear. Sometimes I was on grass when I did this exercise. But sometimes it was on concrete. What were a few bumps, bruises, and grazes? Nothing more than a celebration of my goalkeeping excellence—in my mind's eye, of course. And in that mind were images of last minute heroics at Wembley. "And the youngster Bamforth just made the most unbelievable save, clawing with the very fingertips a screaming shot from Charlton...and Wolves have won! They should all give their medals to this amazing boy goalkeeper. Never have I seen such a display..." This was

the start of my journey—or so I believed—to goalkeeping glory. I would emulate my heroes Malcolm Finlayson, Fred Davies, Phil Parkes, and the rest, and trot onto the park at the Molineux Grounds, Wolverhampton, resplendent in my all-green uniform (I had the kit already as a Christmas present), saving my gold-shirted team time after time with my genius and heroics between the posts. Looking back, there was no doubting my athleticism or bravery. The big shortcoming in the literal sense of the word was my height. I now stand 5 feet 8-and-a bit and was rather shorter then. I spent a lot of time looking at the record books, grateful that two England internationals, by the names of Alan Hodgkinson and Eddie Hopkinson, were hardly giants. The aptly named Steve Death was a keeper of only 5 feet 7 inches. Hope springs eternal. And that is what I would have to do—spring. The second shortcoming was my eyesight. I am profoundly myopic. Since the age of 14 I have worn spectacles every day of my life. The notion of popping contact lenses onto my eyeballs has never appealed, so my goalkeeping career was always bespectacled, usually with a rubber band to secure the glasses to my head. I dreaded rain on match days. Many is the time I lost my specs in the mud. There's a heck of a choice to make when they fall off in the middle of hectic goalmouth action: Sacrifice everything in pursuit of finding and grabbing the ball or ignore the game and retrieve the glasses on which you are dependent to cope in the rest of your daily existence. But, my goodness, perhaps I would have gone for those contacts if I had really and truly done the soccer thing properly.

By now I was playing for the school rugby team—soccer as a sport was not on the curriculum at Up Holland Grammar School—so if I was to develop my career as a footballer, I was going to have to find a team outside of school. Perhaps if I had had a father he would have encouraged this, even insisted upon it, but I didn't. It would be wrong, though, to blame my mother, overprotective as she was of her younger son (no bike for me). We make our own destiny (as I now know), and if I had wanted and believed, I would have been last line for a local team at the very least. Surely I should have made more of the fact that professional goalkeepers were everywhere about me. Our villages outside Wigan cultivated them as productively as the grass that fed Uncle John's Friesian herd. There was David Gaskell, who was Manchester United's last line in the 1963 Cup Final and who had a grocer's shop in Orrell, in which I had been taken out back to see the winner's Wembley medal. There was John Barton, who played for Preston and Blackburn, and who my mother had taught just before he played first team football at the age of 16. Heck, my woodwork teacher at Up Holland Grammar School was Harold Lea, the second-string goalie of Stockport County. Most significant of all, though, was that my mother's cousin by marriage was Wilf Birkett, who lived down the road and who had been a professional goalkeeper with Everton at the time of the last war, and who had later played for Shrewsbury and Southport. He was the trainer (these days they call them physios) at our local team Wigan Athletic. Wigan in those days was a minor league team, but as I write they are in the Premiership (which is a bit like a Class A baseball team making it to the top of the majors). Wilf took me to watch them play and introduced me to the players. What an opportunity—a God-given route to goalkeeping guidance. And still nothing came of it. What was the problem? Was I simply content to daydream?

It was only when I became a PhD student at Hull that I launched into soccer with genuine gusto. Fantasizing was over—this was the real thing. In the intra-varsity leagues there were some good players, some of them semiprofessionals. I could hold my own—I remember in particular one goalless draw where "none shall pass" was writ large across my chest, and there was at lea one notable forward on the opposing side that was utterly frustrated by the guy in the glasses. On the five-a-side circuit we would be watched by the players from the local full-time pro team (Hull City), and I don't recall us doing anything to embarrass ourselves. (I also played a couple of games for one of the reserve teams of Beverley Rugby Union Football Club, the Bulldogs. Then I won the post-game raffle, took the beer home without offering to share, and didn't get on the team sheet ever again.) At Sheffield I continued to play soccer, though with no great frequency, and, because I disliked the all-weather outdoor surface, I strayed out onto the left wing.

When I got to the Brewing Research Foundation, it was to a change of sport—cricket. I had played a bit at school and, as boys, brother John and I had played entire one-a-side cricket games in our yard, in which we would bat and bowl right- or left-handed as we went through entire innings of his Lancashire versus my Worcestershire. Cricket at Nutfield was integral. Lyttel Hall is deep in the Surrey countryside. One thing it had going for it was a lot of green, and what do English people think about when they see an expanse of grass? Why, cricket, of course. Within a week of joining BRF I was lined up for my debut. "Let's see what the new guy can do" was the buzz. I rustled up an old white shirt and a pair of white jeans—but I had never possessed a pair of cricket spikes in my life, so I took to the field for the game (an away game at the local school for the deaf) in trainers. I didn't bowl and I didn't get called on to bat, and my fielding was so-so until the first time I slipped on the damp surface in those smooth-soled pumps. There was much humorous banter at the expense of my profound embarrassment as I dumped myself on my rear end with regularity. No way was I going to repeat the discomfiture, but I was still unconvinced of my long-term commitment to this "new" game. So for the next fixture I emerged from the pavilion wearing my soccer boots. The purists, like lovely Chris Booer with his plummy accent, were appalled, but I didn't slip once and showed just how supple I could be in the field as I brought my goalkeeping dives and safe hands to bear. I was now a regular member of the team, so bought my cricketers boots—and also went to a sports store to buy the most essential part of any cricketer's kit, the "box," which is inserted into the pants to protect the family jewels. I explained to the young female sales assistant (actually my wife's cousin) what I wanted. "Certainly, sir, what size?" I demonstrated that I had the safest pair of hands in the team. But when I got the bat in my hands I became a nervous wreck. You need to know that I don't get nervous about much. I have repeatedly spoken in front of hundreds of people at a time; I have rubbed shoulders comfortably with royalty and superstars; and being crowded on a rugby field or by a gaggle of opposing soccer attackers never fazed me. But the moments before going onto the cricket field to bat for some reason always made me cringe. Sometimes I lived up to the nickname given to me by our best hitter, David Quain, of the "blacksmith," and I would connect to send the ball soaring for fours and sixes (the latter the equivalent of a baseball home run). But even more regularly I would take a mighty swing and completely miss, to hear the sickening smack of leather on wood as the stumps were shattered. If only we had played baseball (a game I have learned to relish since coming to America): At least three chances, and I'm sure I could connect with one of them! Above all, though, cricket is a superbly social game, with one of the key attributes of a good team player being the ability to drink and buy a round in the pub afterwards. In this context, I was a team superstar. I particularly relished games against the brewery sides. Favorite fixture of all would be down in idyllic Dorset, Thomas Hardy country, where we would play in the soft mellow greenery of the Hall and Woodhouse wicket. We would play at Burton against Bass and in Luton. When I joined Bass, they were glad to recruit me into the Research Department team, which played on the ground by the side of the River Trent that was sometimes used by the Derbyshire county side. There were some mightily impressed spectators watching my first at-bat for the team, in which I hit a six and three fours in one inning. Thereafter I don't recall scoring very many runs over five years. I do recall, however, some success with the ball and in the field. I was a reasonable bowler, medium pace with some swing. But it was as a fielder that I excelled, taking enormous delight in swooping low to dive full length and snag catches inches off the ground. After one catch the opposition complained that it was unfair of my team to have roped in a professional fieldsman.

I was persuaded to resurrect my goalkeeping career for the Bass Kelts second team, one of the company sides in the Burton and District League. In the first game I appeared with some hype, but in the fixture against a visiting side from Sheffield I made a complete pig's ear of things, and we were trashed. Three days later I kept a clean sheet and my reputation was restored. Week after week I would glory in the goal, until one dismal Sunday morning on a quagmire of a pitch in the pouring rain (you try to juggle catching a leather football with keeping the raindrops off your spectacles). In the very first attack the tamest of shots was fired at my goal, and I went down like a sack of potatoes, allowing the ball to squirm under my body. As I turned, I watched with agony as it spluttered through the mud and over the line. Our fierce center-half, who humped beer kegs around in his day job, spat that this was the worst bit of goalkeeping he had ever seen in his life, but by the end of the game was fulsome in his praise after I had repelled numerous attacks with the stupid bravery that was my trademark. In the dressing room afterwards, though, I had made my decision, and when I got home that day the boots were thrown into the trash. Only once did I play after that—in a five-a-side tournament that we won, after which I was sent up to collect the trophy, following a display that left me with a bloodied face but happy that I could still deliver between the sticks. And, oh, the joy in one's heart and mind in the bar afterwards, knowing that whatever the size of the ball and whatever its shape, in whatever sport, you had done it for the team.

When I went back to BRF in 1991, it was once more into the cricket team. I was now the head of the research program—and it was my job to recruit new staff. The question about cricket was fairly high on the list: "So what are you? Bowler, batsman?" We appointed two types of people in my sojourn: excellent cricketers and lovely girls. Progressively, though, my suppleness diminished and my place was sacrificed to the younger element. However, a year or so before leaving for America I was persuaded to take part in a Young Sprogs versus Old Farts game at Lyttel Hall. I clearly recall Diane's words to me as I got into the car to drive to the game: "Charlie, no heroics, no diving around. You are not as young as you were." The trouble is, when you have experienced the exhilaration on a football field or a cricket wicket of flying through the air to cleanly catch a ball, it is almost orgasmic in its intensity—or at least extremely adrenaline-filled. What's more, those reflexes aren't lost very readily. Sure enough, on this day a ball was hit in my general direction. I remember thinking "this one is mine," and I took off to my left. Of course I was miles from the ball, but I am sure the dive scored highly for theatrical merit—until I landed. The left arm was extended and the point of contact with the ground was my shoulder. Picture it: The shoulder was on the ground, the left arm pointing towards the sky, and the rest of my body facing downwards with my feet in the air. A sort of shallow "V"—and a pain shooting through my shoulder the likes of which I never experienced before and hopefully never will again. At Diane's insistence I went to the Accident and Emergency unit at Horsham hospital, where they felt around and announced that I had ligament damage. It would need physiotherapy. When I was able to drive again, I started on a weekly course of such treatment. The woman would gently but firmly ease my arm up and down, and I would wince with agony. After several visits, she said that I had better visit a specialist. The doctor in the hospital in Midhurst took a look and pronounced that I had a rotator cuff injury. "It's very common in baseball pitchers. Now here's what we must do: We will give you an MRI and assess the extent of the tear, so that we can see if it can be mended by keyhole surgery or whether we need something a bit more drastic." In due course I found myself being slid into the MRI machine, under instructions to lie perfectly still, and to not be alarmed by the loud banging noises. For a moment or two, things were fine, but then my overactive mind kicked in. "Hmm," I thought, "it must be like this in a coffin." That did it—and before long I was being slid out and given time to breathe deeply so as to come to some sort of equilibrium. It was a situation heaven-sent for meditation, but I hadn't yet found Buddha. Somehow they got me back into the machine and got the images taken. When I was slid out finally, there was a gaggle of medics around. "When you went to the accident room after you did this, what did the X-ray show?" I pointed out that no X-ray was taken. "Pity," they said, "because your humerus is broken. You didn't need this MRI." There, as clear as anything, was a massive crack on the nobbly end that slots into the shoulder. The pints in the first pub I came to in Midhurst were some of the tastiest I have ever had in my life.

2. The Irish question came "very close to home" for me in the shape of Sir Charles Tidbury. Sir Charles was the Chairman of BRF International and someone who engaged very publicly in diverse causes. One of these

was his role as chairman of the William and Mary Ter-
centenary Trust that commemorated the 300th anniver-
sary of the "Glorious Revolution" (1688), in which the
Protestant William of Orange came to the throne. This
put Sir Charles directly into the IRA firing line, and in
June 1990 two Irish Catholics who had attempted to get
into his home were chased to Stonehenge and later
charged with conspiring to murder him. Thereafter Sir
Charles was always accompanied by four armed min-
ders from the Special Branch. To be honest, we at BRF
International felt a sort of vicarious excitement whenever he visited, or when we would visit him for dinners
in London. The visits to us would be preceded by extensive searches by police, and then Sir Charles would
arrive in a car with other cars and bikes before and after.
Guns, walkie-talkies…the whole nine yards. The reality
was that he was very much in danger, and so were we all
through association and proximity. How strange to look
back to realize just how close we were to the problem.

At other times in my life I encountered the problem
firsthand. When I was with Bass, our brewery in Belfast
was in Andersonstown, right in the heart of the conflict.
To be an Englishman, more specifically a Protestant
one, going there truly felt like going into a war zone,
and to be frank, was extremely scary. I had firsthand
knowledge of the fact that money exchanged hands
between the brewery and the local militias to ensure
safe passage of the beer.

Rather earlier in my life, as a university student, we had
a boy, Damian from Londonderry, in my house. I
remember one day clapping suddenly and very loudly
behind him. He nearly leapt out of his skin, so much on
edge was he around loud noises. Londonderry had been
an epicenter of great sectarian strife. He had lived
through it as a child, and to come to university in Eng
land was to escape.

A

The Basics of Malting and Brewing

Fundamentally beer is the product of the alcoholic fermentation by yeast of extracts of malted barley. While malt and yeast contribute substantially to the character of beers, the quality of beer is at least as much a function of the water and, especially, of the hops used in its production.

Barley starch supplies most of the sugars from which the alcohol is derived in the majority of the world's beers. Historically, this is because, unlike other cereals such as wheat, barley retains its husk on threshing, and this husk traditionally formed the filter bed through which the liquid extract of sugars was separated in the brewery.

The starch in barley is enclosed in cell walls and proteins, and these wrappings are stripped away in the malting process (essentially a limited germination of the barley grains), leaving the starch essentially preserved. This softens the grain and makes it more readily milled. Not only that, but unpleasant grainy and astringent characters are removed during *malting*.

Malting

Malting commences with the steeping of barley in water at 14°C to 18°C for up to 48 hours, until it reaches a moisture content of 42 percent to 46 percent. This is usually achieved in at least a three-stage process, with the steeps being interspersed with "air rests" that allow the barley to get some oxygen (to "breathe").

Raising the moisture content allows the grain to germinate, a process that usually takes three to five days at 16°C to 20°C. In germination, the enzymes break down the cell walls and some of the protein in the starchy endosperm, which is the grain's food reserve, rendering the grain friable. Amylases are produced in germination, and these are important for the mashing process in the brewery.

Progressively increasing the temperature during kilning arrests germination, and regimes with progressively increasing temperatures over the range 50°C to perhaps 110°C are used to allow drying to less than 5 percent moisture, while preserving heat-sensitive enzymes. The more intense the kilning process, the darker the malt and the more roasted and burnt are its flavor characteristics.

Brewing

In the brewery, the malted grain must first be milled to produce relatively fine particles, which are for the most part starch. The particles are then intimately mixed with hot water in a process called *mashing*. The water must possess the right mix of salts. For example, fine ales are produced from water with high levels of calcium. Famous pilsners are from water with low levels of calcium. Typically mashes have a thickness of three parts water to one part malt and stand at around 65°C, at which temperature the granules of starch are converted by gelatinization from an indigestible granular state into a "melted" form that is much more susceptible to enzymatic digestion.

The enzymes that break down the starch are called the amylases. They are developed during the malting process, but only start to act once the gelatinization of the starch has occurred in the mash tun. Some brewers will have added starch from other sources, such as corn or rice, to supplement that from malt. These other sources are called adjuncts.

After perhaps an hour of mashing, the liquid portion of the mash, known as wort, is recovered, either by straining through the residual spent grains (*lautering*) or by filtering through plates. The wort is run to the kettle (sometimes known as the copper, even though they are nowadays fabricated from stainless steel) where it is boiled, usually for one hour. *Boiling* serves various functions, including the sterilization of wort, the precipitation of proteins (which would otherwise come out of solution in the finished beer and cause cloudiness), and the driving away of unpleasant grainy characters originating in the barley. Many brewers also add some adjunct sugars at this stage, at which most brewers introduce at least a proportion of their hops.

The hops have two principal components: resins and essential oils. The resins (so-called alpha-acids) are changed ("isomerized") during boiling to yield iso-alpha-acids, which provide the bitterness to beer. This process is rather inefficient. Nowadays, hops are often extracted with liquefied carbon dioxide, and the extract is either added to the kettle or extensively isomerized outside the brewery for addition to the finished beer (thereby avoiding losses due to the bitter substances' tendency to stick onto yeast).

The oils are responsible for the "hoppy nose" on beer. They are very volatile, and if the hops are all added at the start of the boil, then all of the aroma will be blown up the chimney. In traditional lager brewing a proportion of the hops are held back and only added towards the end of boiling, which allows the oils to remain in the wort. For obvious reasons, this process is called late hopping. In traditional ale production, a handful of hops is added to the cask at the end of the process, enabling a complex mixture of oils to give a distinctive character to such products. This is called dry hopping. Liquid carbon dioxide can be used to extract oils as well as resins, and these extracts can also be added late in the process to make modifications to beer flavor.

After the precipitate produced during boiling (trub, which rhymes with *lube*) has been removed, the hopped wort is cooled and pitched with yeast. There are many strains of brewing yeast (*Saccharomyces*), and brewers carefully guard and look after their own strains because of their importance in determining brand identity. Fundamentally, brewing yeast can be divided into ale and lager strains, the former type collecting at the surface of the fermenting wort and the latter settling to the bottom of a *fermentation* (although this differentiation is becoming blurred with modern fermenters). Both types need a little oxygen to trigger their metabolism, but otherwise the alcoholic fermentation is anaerobic. Ale fermentations are usually complete within a few days at temperatures as high as 20°C, whereas lager fermentations at as low as 6°C can take several weeks. Fermentation is complete when the desired alcohol content has been reached and when an unpleasant butterscotch flavor which develops during all fermentation has been mopped up by yeast. The yeast is harvested for use in the next fermentation.

In traditional ale brewing the beer is now mixed with hops, some priming sugars, and with isinglass finings from the swim bladders of certain fish, which settle out the solids in the cask.

In traditional lager brewing the "green beer" is matured by several weeks of cold storage, prior to filtering.

Nowadays, the majority of beers, both ales and lagers, receive a relatively short conditioning period after fermentation and before filtration. This conditioning is ideally performed at –1°C to –2°C for a minimum of three days, under which conditions more proteins drop out of the solution, making the beer less likely to go cloudy in the package or glass.

The filtered beer is adjusted to the required carbonation before packaging into cans, kegs, or glass or plastic bottles.

Types of Beer

Type of Beer	Origin	Typical Range of Alcoholic Strength (% by vol.)	Characteristics
Ales & Stouts			
Bitter (Pale) Ale	England	3-5.5	Dry hop, bitter, estery, malty, low carbonation, copper color.
India Pale Ale	England	5-7	Bitter, hoppy.
Scottish Ale	Scotland	3-3.5 ("light"); 3.5-4 ("heavy")	Malty, sweet.
Old Ale	England	6-9	Dark amber-brown; malty, sweet.
Alt (n. b. Alt means "old")	Germany	4+	Some esters, bitter, copper color.
Red Ale	Ireland	4-4.5	Sweet, malty, somewhat bitter.
Oud Bruin and Oud Red Ales	Belgium	4.8-5.2	Fruity, sour, some maltiness, woody.
Mild (brown) Ale	England	<3.5	Dark brown, sweet, mellow.
Stout	Ireland	4-5	Roast, bitter, black.
Export Stout	Ireland	6-9	Roast, bitter, black.
Imperial Stout	England	7-12	Dark, very malty, fruity, quite bitter but not very hoppy.
Porter	England	4.5-6.5	Similar to stout but less roast character.

Type of Beer	Origin	Typical Range of Alcoholic Strength (% by vol.)	Characteristics
Ales & Stouts			
Sweet Stout	England	3.5-4	Sweet, dark brown/black, not very roasty.
Oatmeal Stout	England	4-6	Full, dry, some roast.
Barley Wine	Britain	8-10	Estery, copper/brown.
Kölsch	Köln (Cologne) Germany	4.4-5	Pale gold, light and dry (ale/lager hybrid).
Dubbel	Belgium	6-7.5	Red to dark brown, malty, medium bitterness.
Tripel	Belgium	7-10	Fruity, spicy, medium bitterness.
Table Beer	Belgium	0.5-3.5	Light bodied, low carbonation, low bitterness.
Saison	Belgium	4.5-8.5	Golden to deep amber, very fruity/estery, reasonable bitterness, may be spicy.
Weizenbier (wheat beer)	Germany	5-6	Cloves, slightly cloudy, straw color.
Weisse	Germany	2.5-3.5	Pale, sour.
Wit	Belgium	4.8-5.2	Pale, spiced with coriander and orange peel.
Lambic	Belgium	5-7	Amber, often cloudy, fruity, sour; therefore often incorporates fruit—cherries (kriek), raspberries (framboise), peaches (peche), black currants (cassis), or apples (pomme).

Type of Beer	Origin	Typical Range of Alcoholic Strength (% by vol.)	Characteristics
Lagers			
Pilsner/Pils	Czech Republic	4-5	Late hop, full-bodied, malty, pale amber/gold.
Bock	Germany	6-8	Sulfur, malty, colors ranging from straw (Maibock) to dark brown (Doppelbock). Eisbocks are even stronger (8.5-15%) due to an ice-forming stage post-fermentation.
Helles	Germany	4.5-5.5	Pale amber/gold, very malty, low bitter/hop character.
Märzen (meaning "March" for when traditionally brewed)	Germany	5-6	Medium bitter/hop; toasted character; amber through reddish brown. The Vienna style is very similar.
Dunkel	Germany	4.5-5	Copper-brown, malt/toast.
Schwarzbier	Germany	3.8-5	Toast, caramel, dry, black.
Malt liquor	United States	6.25-7.5	Malt/sweet, little hop, alcoholic, pale. Many states decree that any beer containing more than 5.5% ABV must be so declared.
Rauchbier	Germany	4.5-6.5	Smoked malt, amber/brown.

The Brewers Association issues extensive guidelines for these and other beers, listing inter alia as extra categories the beers developed by the craft sector in the United States that are founded on the historical beers from other nations. Go to http://www.beertown.org/education/pdf/BA_Beer_Style_2009.pdf.

About the Author

Charles W. Bamforth has been in the brewing industry for 32 years, including 13 years in research, 11 in academia, and 8 with the famed brewing company Bass. After an international search, he was selected as UC Davis' first Anheuser-Busch Endowed Professor of Malting and Brewing Sciences. His Web site gives fuller details of his career and much interesting information about beer and brewing, besides: http://www-foodsci.ucdavis.edu/bamforth/.

Throughout his diverse career, he has embraced every dimension of beer, from raw materials and processing, through quality, to beer's impact on the body. This makes him unique among "beer people" worldwide. He has published many research papers in the peer-reviewed domain, but also those targeted at the layperson, seeking to engage awareness and debate about beer as a product and as part of social fabric. This is his ninth book on beer (one of his earlier ones is in its third edition), and he is generally considered to be one of the world's leading writers and speakers on beer, from an authoritative, but also humorous and engaging, perspective. In recent years his major research thrust has been on the wholesomeness and public perception of beer.

INDEX

A

Abbot Ale, 157
acetaldehyde, 110
aerobic water treatment, 168
Aggies, 137
Alcohol Flush Reaction, 165
alcohol-free beer, 39-42
Alcohol: Its Action on the Human Organism (D'Abernon), 103
alcoholic strength of beers, table of, 219-221
alcopops, 95
aldehyde dehydrogenase enzyme, 165
ales
brewing, 216-217
types of, 219-220
alpha-acids, 215
alt ale, 219
Alzheimer's Disease, reduced risk with moderate beer consumption, 109
Amalgamated Union of Engineering Workers, 24
AmBev (Companhia de Bebidas das Américas)
acquisition by Interbrew, 8
history of, 5
American Bar Association, resolution for repeal of prohibition, 89

American Homebrewers Association, 62-63
American Society of Brewing Chemists, 171
American Temperance Union (ATU), 87
amylases, 215
anaerobic water treatment, 168
Anchor Brewing Company, 54-58
Anderson, Ray, 203
Anheuser-Busch. *See also* Anheuser-Busch InBev
acquisition by InBev, 4
acquisition of Rolling Rock brand, 18
donation to University of California, Davis, 136
investments, 4, 137
quality control, 1-4
Anheuser-Busch InBev, xvi, 34
brand rationalization, 16-19
closure of Stag Brewery (London), 8-9
formation of, 4
size of, 137
Antarctica Paulista, 5-6
anti-alcohol forces
anti-alcohol organizations in U.S., 90-91, 186
Mothers against Drunk Driving (MADD), 89-90
prohibition, 88-89

Vice President, Publisher: Tim Moore
Associate Publisher and Director of Marketing: Amy Neidlinger
Acquisitions Editor: Kirk Jensen
Editorial Assistant: Pamela Boland
Operations Manager: Gina Kanouse
Senior Marketing Manager: Julie Phifer
Publicity Manager: Laura Czaja
Assistant Marketing Manager: Megan Colvin
Cover Designer: Alan Clements
Managing Editor: Kristy Hart
Project Editors: Jovana San Nicolas-Shirley and Kelly Craig
Copy Editor: Geneil Breeze
Proofreader: Seth Kerney
Indexer: Erika Millen
Senior Compositor: Gloria Schurick
Manufacturing Buyer: Dan Uhrig

Publishing as FT Press

Upper Saddle River, New Jersey 07458

FT Press offers excellent discounts on this book when ordered in quantity for bulk purchases or special sales. For more information, please contact U.S. Corporate and Government Sales, 1-800-382-3419, corpsales@pearsontechgroup.com. For sales outside the U.S., please contact International Sales at international@pearson.com.

Company and product names mentioned herein are the trademarks or registered trademarks of their respective owners.

Printed in the United States of America

First Printing September 2010

ISBN-10: 0-13-706507-8
ISBN-13: 978-0-13-706507-3

Pearson Education LTD.
Pearson Education Australia PTY, Limited.
Pearson Education Singapore, Pte. Ltd.
Pearson Education Asia, Ltd.
Pearson Education Canada, Ltd.
Pearson Educación de Mexico, S.A. de C.V.
Pearson Education—Japan
Pearson Education Malaysia, Pte. Ltd.

Library of Congress Cataloging-in-Publication Data

Bamforth, Charles.

Beer is proof God loves us : reaching for the soul of beer and brewing / Charles Bamforth. — 1st ed.

p. cm.

ISBN 978-0-13-706507-3 (hardcover : alk. paper) 1. Beer—History. 2. Brewing industry—History. I. Title.

HD9397.A2B36 2010

338.4'766342—dc22

2010023040

Beer Is Proof
God Loves Us

Reaching for the Soul of Beer and Brewing

Charles W. Bamforth